REFLEXIVE EPISTEMOLOGY

BOSTON STUDIES IN THE PHILOSOPHY OF SCIENCE

VOLUME 118

DANILO ZOLO

Department for Philosophy and Social Sciences
University of Siena, Italy

REFLEXIVE EPISTEMOLOGY

The Philosophical Legacy of Otto Neurath

Translated from the Italian by David McKie

KLUWER ACADEMIC PUBLISHERS
DORDRECHT / BOSTON / LONDON

Library of Congress Cataloging-in-Publication Data

Zolo, Danilo.
[Sciénza e politica in Otto Neurath. English]
Reflexive epistemology : the philosophical legacy of Otto Neurath / Danilo Zolo ; [translated from the Italian by David McKie and Danilo Zolo].
p. cm. -- (Boston studies in the philosophy of science ; v. 118)
Translation of: Scienza e politica in Otto Neurath.
Includes bibliographies and indexes.
ISBN 0-7923-0320-2 (U.S. : alk. paper)
1. Neurath, Otto, 1882-1945. I. Title. II. Series.
Q174.B67 vol. 118
[B3309.N394]
001'.01 s--dc20
[193] 89-11171

ISBN 0–7923–0320–2

Prima Edizione in "Campi del sapere" gennaio 1986

Published by Kluwer Academic Publishers,
P.O. Box 17, 3300 AA Dordrecht, The Netherlands.

Kluwer Academic Publishers incorporates
the publishing programmes of
Martinus Nijhoff, Dr W. Junk, D. Reidel, and MTP Press.

Sold and distributed in the U.S.A. and Canada
by Kluwer Academic Publishers,
101 Philip Drive, Norwell, MA 02061, U.S.A.

In all other countries, sold and distributed
by Kluwer Academic Publishers Group,
P.O. Box 322, 3300 AH Dordrecht, The Netherlands.

Printed on acid-free paper

Printed in the Netherlands

Physical theory [...] secures its value not from another method which, applied at the same time to the same objects, would make up for the insufficiencies of the physical method and would confer on its theories a value transcending their own nature. There is no method except the physical method which can serve to study the objects studied by physics; the physical method in itself exhausts the justification of physical theories; it and it alone indicates what these theories are worth as knowledge. [...] Science is impotent to establish the legitimacy of the principles themselves which outline its methods and guide its researches. [...] At the bottom of our most clearly formulated and most rigorous doctrines we always find again [a] confused collection of tendencies, aspirations, and intuitions. No analysis is penetrating enough to separate them or to decompose them into simpler elements. No language is precise enough and flexible enough to define and formulate them.

[P. Duhem, *The Aim and Structure of Physical Theory*]

Duhem has shown with special emphasis that every statement about any happening is saturated with hypotheses of all sorts and that these in the end are derived from our whole world-view. We are like sailors who on the open sea must reconstruct their ship but are never able to start afresh from the bottom. [...] They make use of some drifting timber of the old structure, to modify the skeleton and the hull of their vessel. But they cannot put the ship in dock in order to start from scratch. During their work they stay on the old structure and deal with heavy gales and thundering waves. [...] That is our fate.

[O. Neurath, *Anti-Spengler*; *Foundations of the Social Sciences*]

The philosopher's task was well compared by Neurath to that of a mariner who must rebuild his ship on the open sea. We can improve our conceptual scheme, our philosophy, bit by bit while continuing to depend on it for support; but we cannot detach ourselves from it and compare it objectively with an unconceptualized reality. [...] Epistemologists have wanted to posit a realm of sense data [...] for fear of circularity [...] But with Neurath we accept circularity, simply recognizing that the science of science is a science.

[W.V.O. Quine, *From a Logical Point of View*; *The Ways of Paradox*]

CONTENTS

EDITORIAL PREFACE

Professor Danilo Zolo has written an account of Otto Neurath's epistemology which deserves careful reading by all who have studied the development of 20th century philosophy of science. Here we see the philosophical Neurath in his mature states of mind, the vigorous critic, the scientific Utopian, the pragmatic realist, the sociologist of physics and of language, the unifier and encyclopedist, always the empiricist and always the conscience of the Vienna Circle. Zolo has caught the message of Neurath's ship-at-sea in the reflexivity of language, and he has sensibly explicated the persisting threat posed by consistent conventionalism. And then Zolo beautifully articulates the provocative theme of the 'epistemological priority of sociology'.

Was Neurath correct? Did he have his finger on the pulse of empiricism in the time of a genuine unity of the sciences? His friends and colleagues were unable to follow all the way with him, but Danilo Zolo has done so in this stimulating investigation of what he tellingly calls Otto Neurath's 'philosophical legacy'.

R. S. COHEN

ABBREVIATIONS

'Pseudo' = [Otto Neurath], 'Pseudorationalismus der Falsifikation', *Erkenntnis*, *5* (1935), pp. 353–65.

Foundations = [Otto Neurath], *Foundations of the Social Sciences*, in *International Encyclopedia of Unified Science*, vol. 2, no. 1, pp. 1–51, Chicago: The University of Chicago Press, 1944.

ES = Otto Neurath, *Empiricism and Sociology*, ed. by M. Neurath and R.S. Cohen, Dordrecht and Boston: D. Reidel, 1973.

GpmS = Otto Neurath, *Gesammelte philosophische und methodologische Schriften*, ed. by R. Haller and H. Rutte, Wien: Hölder-Pichler-Tempsky Verlag, 1981, vol. I and vol. II.

PP = Otto Neurath, *Philosophical Papers 1913–1946*, ed. by R.S. Cohen and M. Neurath, Dordrecht and Boston: D. Reidel, 1983.

ACKNOWLEDGMENTS

Among the many people who have aided my research on this book my thanks are due above all to Robert S. Cohen, Chairman of the Center for Philosophy and History of Science at Boston University, who invited me as Research Associate for the academic year 1981-82 and enabled me to participate in the work of the Boston Colloquium for the Philosophy of Science and in his own seminars on the Vienna Circle. It was from his invitation to prepare a study on Neurath for the *Boston Studies in the Philosophy of Science* that the original idea of this book sprang, and to him also I owe the opportunity of publishing my work in this present English edition.

As a result of the hospitality of Boston University I have been able to discuss on a number of occasions the general themes and structure of my work with Joseph Agassi, Carl G. Hempel, Thomas A. McCarthy, Ernest Nagel and Marx W. Wartofsky, all of whom I wish to take this opportunity of thanking. In addition my bibliographical searches at Boston were greatly facilitated by the kind assistance of Alessandro Pizzorno, who also helped to introduce me to the resources of the Widener Library at Harvard University.

Another major debt is due to Michael A. Hoskin, Head of the Department of History and Philosophy of Science at the University of Cambridge, and to Mary Hesse, who welcomed me to their Department as Visiting Scholar during the summer and autumn of 1986. My stay at Cambridge provided me with ideal conditions for preparing the new material for this English version of my work and I was stimulated by a number of interesting conversations with Mary Hesse to recast several sections in the Introduction and Conclusion. In this I was greatly helped by the small but select collection of the Department's own library and by the splendid riches of the Cambridge University Library.

I am grateful also for the particularly kind hospitality shown to me by Nicholas Rescher, Director of the Center for Philosophy of Science at Pittsburgh University, Pennsylvania, who invited me as Senior Visiting Fellow for the first semester of the academic year 1987-88. This second American visit has allowed me to complete my bibliographical work in the Hillman Library, where I have consulted, among much else, the *Carnap Collection* which includes an important exchange of letters between Neurath and Carnap. The final pages of this book were written in my room on the eighth floor of the University's famous Cathedral of Learning, and a Lunchtime Talk organised by the Center for Philosophy of Science gave me the chance to present the basic ideas behind this book and to discuss them with, in particular, my colleagues William Bechtel, Paul Hoyningen-Huene, Peter Janich, Ulrich Majer, Nicholas Maxwell and Victor Rodriguez.

Of the numerous Italian friends who have discussed this book with me during the time between its original Italian edition and publication of the present English version, I wish to record my special gratitude, for their criticism as much as for their agreement, to Mario Caciagli, Furio Cerutti, Pietro Costa, Luigi Ferrajoli, Riccardo Guastini, and Giovan Francesco Lanzara. I am further grateful to Paolo Rossi for his most helpful observations on the history of science and to Sergio Moravia for the valuable interest which he has on several occasions shown towards my work. I should add that, but for the initial encouragement of Norberto Bobbio, it is entirely likely that I should never have felt able to undertake the writing of this book.

My thanks are due also to Rudolf Haller, of the Karl-Franzens-Universität at Graz, for welcome bibliographical assistance at Vienna and Graz, and to Claus Offe for bringing several useful works to my attention; also to Cristina Bicchieri, who has several times now eased the path for me to university departments in England and in America; to Alessandra Maccioni, who helped me in Cambridge to draft the bibliography which accompanies the English text; to Elena Esposito, who carried out for me further complicated bibliographical research at the University of Bielefeld; and finally to David McKie, Fellow of Robinson College, Cambridge, who has undertaken the delicate and onerous task of creating for me an English version removed from the flatness of the 'standard English' to which, without his thoughtful cooperation, I fear I should otherwise have had to resign myself.

Cambridge, October 1988 D. ZOLO

INTRODUCTION

The purpose of this book is to present an examination of the thought of Otto Neurath together with an analysis of the historical development of a number of central aspects in mid-European logical positivism. Since, however, a new interpretation of Neurath's thought can hardly be formed without the attempt to offer arguments in support of certain more general theses, it is, whether successful or not, in this attempt that I would like to see the interest and importance of my work. With this in mind, therefore, the book has been written and will, I hope, prove capable of being read in the same light. For this reason, and because reference to my more general theses cannot be made explicit at every point during the reconstruction of Neurath's thought, it may be found useful for those theses to be sketched in preliminary outline here.

From both the historiographical and the philosophical position the overall theses to which I subscribe may be seen encapsulated within the three successive quotations of Duhem, Neurath and Quine, with which the book opens and at whose centre stands the famous nautical metaphor formulated by Neurath.

On its most basic level the historiographical thesis which I intend to maintain may be put as follows: that the thought of Neurath represents, to a far greater extent than tends to be recognised, a significant anticipation of the (self-)criticism of logical positivism and of the 'received view' of empiricism which Quine has advanced in such works of undoubted philosophical importance as, amongst others, 'Two Dogmas of Empiricism' and 'Epistemology Naturalized'. Neurath, we shall find, was the only writer among those active in the first decades of this century to have retained, from a position within European logical positivism, a strict and unambiguous connection with the classical tradition of French conventionalism, especially with the 'holistic thesis' of Duhem and the cognitive relativism of Poincaré and Rey.

Put more broadly, and in all probability more controversially, this historiographical thesis may be stated as follows: that the 'revolt against positivism' promoted in Anglo-American circles in the Seventies by neo-Popperian, post-Popperian and, more generally, by post-empiricist writers was, in point of fact, a non-existent 'revolt'. Without in any way setting out to deny the merits of such writers as Hanson, Scriven, Toulmin, Agassi, Lakatos, Feyerabend or Quine himself and without intending to diminish the importance of a book such as Thomas Kuhn's *The Structure of Scientific Revolutions*,[1] it is entirely possible to argue that the basic impulses behind the antipositivist revolt of the Seventies were neither new nor revolutionary.

In truth all of these impulses were already present within European – that is to say Austrian, German, Polish and also Italian – philosophy of science of the early years of this century. Most importantly they were present, in opposition to the orthodoxy of Schlick and Carnap, at the centre of the thought of the very Vienna Circle which the majority of people, following Popper, see as a hotbed of positivistic dogmatism. As Philipp Frank has shown, the central ideas of the conventionalistic critique of positivism lay already at the heart of the 'first' Vienna Circle in the years 1908–12. The 'discussions in a Vienna Coffee House' between Neurath, Frank and Hahn were strongly influenced by the conventionalism of Abel Rey, whose book, *La théorie physique chez les physiciens contemporains*, had immediately been translated into German in 1908 and was constantly the subject of their discussions. The idea which it made central to the concerns and later convictions of the 'left wing' of the Circle was that of the 'crisis of science', that is to say the decline of mechanistic physics and the positivistic metaphysics which they entailed.[2]

It is these same ideas which are to be seen at work in the writings of Ludwik Fleck, whose contribution to the thought of Thomas Kuhn is probably a great deal more important than is generally seen and perhaps even more so than Kuhn himself has yet recognised. Furthermore, Fleck's thought, as indeed Neurath's also, came under the direct influence of Kazimierz Ajdukiewicz, one of the leading figures of the Lvov-Warsaw school and also an active collaborator on *Erkenntnis*. Ajdukiewicz's 'radical conventionalism' was in reality an explicit and coherent development of classical French conventionalism, especially of the thought of Poincaré and Le Roy.[3] And it is this which helps to explain a number of otherwise surprising affinities between the epistemologies of Neurath and Fleck, especially between Neurath's notion of '*Enzyklopädie*' and Fleck's of '*kolektyw myślowy*' (thought-collective).

Also significant in this context is the Italian tradition of mathematics and philosophy of science, which took for the most part a pragmatistic and conventionalistic direction. A book such as Federigo Enriques' *I problemi della scienza*,[4] for example, exercised considerable influence on Neurath. In the course of his work Neurath frequently referred to Enriques' ideas and took over the fears several times expressed by him that within the

'*Logizismus*' of Viennese orthodoxy there lay concealed grave dangers of dogmatism and scholasticism.

One important corollary of this historiographical thesis appears to me to be the refutation of the idea that Popperian falsificationism represents, in any kind of accepted meaning of the term, a 'liberalisation' of Viennese logical positivism. While Neurath, Frank and Hahn within the Vienna Circle received the central ideas behind classical conventionalism and proved themselves alert to the indeterministic and probabilistic implications of quantum physics, Popperian falsificationism grew out of direct conflict with conventionalistic ideas and the uncertainty principle of the Copenhagen school.

From Popper's point of view Neurath's epistemological and sociological relativism was a typical example of the 'poverty of historicism'. But, from the point of view of Neurath's critique of 'pseudorationalism', Popper's philosophy was a 'fallibilist' variant of traditional realistic metaphysics, as Neurath maintained from 1935 on and as he never tired of reiterating in his sharp epistolary interchanges with Carnap in the last years of his life. And this was something finally acknowledged by Popper himself:

> Neurath speaks quite correctly of my 'metaphysical tendencies' [...He] realised that I really was not a positivist and, in some sense of the word, not even an empiricist [...]. I am a realist.[5]

Strongly linked, however, with these 'irrational' metaphysical tendencies were the premises behind Popper's political philosophy and his celebrated critique of Marxism.

The philosophical thesis which I should like to see advanced by my work concerns above all the depth of meaning which I believe to be contained within Neurath's nautical metaphor. There he likened the condition of scientists to that of sailors in mid-ocean who are forced to repair and restructure their ship while the storm rages about them. They are unable to return to port and build the ship anew in dry dock. Consequently, during the act itself of alteration and repair, they are forced to support themselves on the very structure which stands in need of repair. This metaphor has undoubtedly exercised a deep influence on Quine's thought.[6] At the same time Quine's frequent quotation of it has further increased both its notoriety and its influence. Following him, it has been used in various ways and in various contexts, especially by such philosophers and epistemologists as P. Lorenzen,[7] H. Blumenberg,[8] and C. Cherniak.[9] In certain not insignificant ways, however, Quine's interpretation is a reductive one. Quine, for his purposes, sees Neurath's metaphor as signifying an 'empiricism without dogmas' which has "stopped dreaming of deducing science from sense data". He accepts fully the circularity of an epistemological position which no longer makes reference to extralinguistic data on which to base the propositions of science, but turns it all the same towards support of a naturalistic reduction of epistemology. The insidious threat of circularity is removed, he argues, if, once the circularity is recognised and accepted for what it is, metalinguistic (epistemological) analysis of the processes of empirical research is made to coincide with its language-object, "simply recognizing that science of science is a science". Recognition of the condition of circularity in epistemology becomes therefore the premise for an anti-sceptical and anti-philosophical conclusion. If consistently practised, Quine suggests, the Cartesian method leads us into a cognitive paralysis: we cannot meaningfully question the correctness of our total scheme.[10] With 'philosophical' epistemology removed, along with positions of methodical doubt (i.e. any doubt not justified on the basis of specific reasons) he then proposes the construction of a 'naturalised epistemology':

> The old epistemology aspired to contain natural science; [...]. Epistemology in its new setting, conversely, is contained in natural science, as a chapter of psychology. [...] We are after an understanding of science as an institution or process in the world, and we do not intend that understanding to be any better than the science which is its object. This attitude is indeeed one that Neurath was already urging in Vienna Circle days, with his parable of the mariner who has to rebuild his boat while staying afloat in it.[11]

It is doubtful, in my view, that this is exactly what Neurath had in mind. But it is above all doubtful that reducing epistemology to naturalistic psychology can offer an answer to the classical 'philosophical' problems of epistemology, and does not, instead, as Richard Rorty has maintained, amount to a quasi-positivistic negation of those problems.[12] Quine's formulation of a 'naturalised epistemology' does, however, seem to me to be a premise entirely in accord with epistemological conceptions which tend today in different ways to consider epistemology as a branch of cognitive science. See for instance D.T. Campbell's 'evolutionary epistemology',[13] and W.S. McCulloch and H. von Foerster's 'experimental epistemology'.[14] And consider in addition B. Barnes and D. Bloor's 'strong programme'.[15] Such epistemologies as these react to the situation of circularity and of the absence of foundations with a reductive, polemically non-philosophical, concept of 'reflexivity', intending this to mean, in Quinean fashion, the immediate recursive applicability of empirical methods and results to epistemology, the theory of knowledge, or, as in the many ways typical case of the 'Edinburgh school', to the sociology of knowledge and the sociology of science.

My own interpretation of Neurath's metaphor is, as will be seen, both radical and 'philosophical' and is one which sets out to reveal a deep and important vein in Neurath's thought, running close to his critique of Cartesian 'pseudorationalism'. The metaphor suggests, I maintain, the idea of a general and insuperable situation of linguistic and conceptual circularity which prevents any attempt at justification or 'directional' self-foundation in scientific knowledge and which rules out any 'constructive philosophy' aiming to break circularity of thought or of language in pursuit of some 'methodological beginning'.[16] Such a position of insurmountable circularity involves the recursivity of the language of science, self-inclusion of the foundations of mathematics and formalised logical systems, self-reference of the methods and tests of physics, and holistic irreversibility of highly specialised developments within individual areas of empirical research.

The only possible reply to this situation of circularity is reflexive consciousness of the circularity and of the absence of foundations. This consciousness has undoubtedly dangerous and paradoxical sides to it in that it dwells in the difficult position between the threat on the one hand of cognitive paralysis and aphasia – Descartes' methodical doubt and the fruitless search for a *tabula rasa* – and on the other the realisation of the at once both dogmatic and precarious nature of every advance in scientific discourse inasmuch as it is cognitive discourse influenced by *tabulae inscriptae* which are either hidden or indecipherable.

Reflexive recognition of the situation of circularity is the very task of philosophy, conceived not as an autonomous discipline, as a special kind of knowledge about knowledge, but as a permanent self-criticism of the historical forms of theoretical and practical knowledge. This recognition demands a disposition towards the stipulative renegotiation of one's points of departure and, in addition, the honest admission that every scientific convention, however analytical, formally refined, or reflexive it may be, still contains elements which are non-rational and non-analysable. These are the elements involved in those "confused collections of tendencies, aspirations, and intuitions" of which Duhem spoke and which form part of the 'folklore' (or the 'metaphysics') of each historical period, culture, and social group, from which no individual philosopher or scientist or scientific community has the capacity to be detached. If it is impossible therefore to separate the language of science from the language of such things as emotions, hopes, fears, intuitions, collective prejudices, traditions, and authority, then the positivistic and behaviouristic dogma of *Wertfreiheit* will have to be abandoned. Also the attempts will have to be rejected of such recent writers as John Rawls, J.C. Harsanyi, Robert Nozick and Donald Dworkin to renew the old European tradition of ethical cognitivism and political moralism by appeal to the validity of ontologically presupposed or discursively argued universal ethical principles. And, in epistemology, any idea of the hegemony of the natural as opposed to the human or social sciences will have to be abandoned. With Quine we might still be able to continue to use the term 'science' for this 'science of science', but it will not in any way be capable of being understood or practised as a 'natural science' without becoming both 'pseudo-epistemology' and at the same time, as Laudan suggests with reference to the Edinburgh school, 'pseudo-science'.[17] Mary Hesse correctly argues in a recent work that, rather than 'naturalising' epistemology, it is more a question of 'socialising' and 'historicising' it.[18] Instead of trying to make epistemology a part of natural science, it is necessary to recognise the epistemological priority of the social sciences and, following Fleck's suggestion, of the interpretative and historiographical approach to the study of science.[19] No 'rational' description of the growth of the empirical sciences can possibly dispense with the semantic, historical, and sociological approach to the behaviour, languages, shared values, techniques of persuasion, and methodological decisions of scientific communities.

It seems necessary therefore to recognise (along with Neurath, but beyond Quine and in opposition to naturalised epistemologies of the Barnes and Bloor type) that epistemological research can only start from the reflexive interpretation of its own symbolic universe – a philosophical operation *par excellence* which no naturalistic short circuit can supplant.[20] Obviously this will only hold true so long no intention exists to strip epistemology of its essential responsibility for questions on the 'meaning', methods and goals of scientific knowledge as such. It will be true also if, in this age of nuclear power, robotics, and information technology, it still makes sense to ask whether modern science is not only the producer of technological development and an increasing ability to manipulate our environment but is also, in a not wholly instrumental sense, a worthwhile and useful form of knowledge.

NOTES

[1] It is perhaps worth mentioning that this book originally appeared as a monograph in the second volume of the *International Encyclopedia of Unified Science*, which had been founded by Neurath, and immediately following Neurath's essay *Foundations of the Social Sciences*, which (notwithstanding Carnap's opposition) opened the volume. Was this mere coincidence?

[2] Cf. P. Frank, *Modern Science and Its Philosophy*, Cambridge: Harvard University Press , 1950, pp. 1–4. See also: P. Rossi, 'Introduzione' to L. Fleck, *Nascita e sviluppo di un fatto scientifico*, Bologna: Il Mulino, 1983, p. 31.

[3] Cf. J. Giedymin, 'Polish Philosophy in the Inter-War Period and Ludwik Fleck's theory of Thought-Styles and Thought-Collective' in R.S. Cohen and T. Schnelle (eds.), *Cognition and Fact. Materials on Ludwik Fleck*, Dordrecht and Boston: D. Reidel, 1986, pp. 179–215. On this subject see also the contributions of B. Wolniewicz, W. Markiewicz and T. Schnelle, ibid., pp. 217–21, 223–9, 231–65.

[4] F. Enriques, *I problemi della scienza*, Bologna: Zanichelli, 1906, English trans. *Problems of Science*, Chicago and London: Open Court, 1914.

[5] K.R. Popper, 'Replies to My Critics', in P.A. Schilpp (ed.), *The Philosophy of Karl Popper*, La Salle (Ill.): Open Court, 1974, pp. 1183, 971. Popper has recently repeated against Kuhn his view that, compared with physics, human sciences such as psychology and sociology are "spurious sciences" which "are riddled with fashions, and with uncontrolled dogmas" (K.R. Popper, 'Normal Science and Its Dangers', in I. Lakatos and A. Musgrave, eds., *Criticism and the Growth of Knowledge*, Cambridge: Cambridge University Press, 1970, pp. 58, 57–8).

[6] Cf. G.D. Romanos, *Quine and Analytic Philosophy*, Cambridge (Mass.): MIT Press, 1983, pp. 21–3, 105.

[7] P. Lorenzen has referred to Neurath's metaphor in *Methodisches Denken*, Frankfurt a.M.: Suhrkamp Verlag, 1968, pp. 24–59, English trans. in *Constructive Philosophy*, Amherst: The University of Massachusetts Press, 1987, pp. 3–29. ("If we envision natural language as a ship at sea, then our situation can be described as follows: if we are unable to make landfall, then our ship must have been constructed on the high sea – not by us but by our ancestors. Our ancestors must have been able to swim and have somehow carpentered together a raft out of, say, driftwood. They then continually improved on this raft until today the raft has become a comfortable ship. So comfortable that we no longer have the courage to jump into the water and once more start from scratch. To solve the problem of the method for thought, we must put ourselves in such a shipless condition, that is, bereft of language, and then attempt to retrace the activities whereby we could, while swimming free in the middle of the sea of life, build for ourselves a raft or even a ship", ibid., p. 6).

[8] H. Blumenberg has interpreted Neurath's metaphor as a circular development of Lucretius' metaphor of 'shipwreck with a witness' (*Schiffbruch mit Zuschauer. Paradigma einer Daseinsmetapher*, Frankfurt a.M.: Suhrkamp Verlag, 1979, chap. 6, passim).

[9] Cf. C. Cherniak, *Minimal Rationality*, Cambridge (Ma): The MIT Press, 1986, pp. 107–9, 120, 123–5, 130.

[10] Cf., ibid., p. 108.

[11] W.V.O. Quine, 'Epistemology Naturalized', in W.V.O. Quine, *Ontological Relativity and Other Essays*, New York and London: Columbia University Press, 1969, pp. 83–4.
[12] Cf. R. Rorty, *Philosophy and the Mirror of Nature*, Oxford: Basil Blackwell, 1980, pp. 221–30; see also, ibid., pp. 165–73.
[13] Cf. D.T. Campbell, 'Evolutionary Epistemology', in P.A. Schilpp (ed.), *The Philosophy of Karl Popper*, pp. 413–63. See also: W. Callebaut and R. Pinxten (eds.), *Evolutionary Epistemology. A Multiparadigm Program*, Dordrecht and Boston: D. Reidel, 1987; M. Ruse, *Taking Darwin Seriously. A Naturalistic Approach to Philosophy*, Oxford: Basil Blackwell, 1986, particularly pp. 29–66.
[14] Cf. H. Von Foerster, *Observing Systems*, Seaside (Cal.): Intersystems Publications, 1984.
[15] Cf. B. Barnes and D. Bloor, "Relativism, Rationalism and the Sociology of Knowledge", in M. Hollis and S. Lukes (eds.), *Rationality and Relativism*, Oxford: Basil Blackwell, 1982, pp. 20–47. See also: B. Barnes and D. Edge (eds.), *Science in Context*, Milton Keynes: The Open University Press, 1982.
[16] Cf. P. Lorenzen, *Constructive Philosophy*, pp. 5–6.
[17] L. Laudan, 'The Pseudo-science of Science?', in J.R. Brown (ed.), *Scientific Rationality: The Sociological Turn*, Dordrecht and Boston: D. Reidel, 1984, pp. 41–73.
[18] M. Hesse, 'Socializing Epistemology', in E. McMullin (ed.), *Construction and Constraint,* Notre Dame: University of Notre Dame Press, 1988, p. 98. See also M. Hesse, *Revolutions and Reconstructions in the Philosophy of Science*, Brighton: The Harvester Press, 1980, pp. 29–60, with an explicit reference to the 'Neurath-Quine metaphor' (p. 41).
[19] Cf.: L. Fleck, 'Problems of the Science of Science', in R.S. Cohen and T. Schnelle (eds.), *Cognition and Fact. Materials on Ludwik Fleck*, pp. 113–27; L. Fleck, 'The Problem of Epistemology', ibid., pp.79–112.
[20] On this see: S.J. Bartlett and P. Suber (eds.), *Self-Reference*, Dordrecht: Martinus Nijhoff Publishers, 1987; E.R. Fuhrman and K. Oehler, 'Discourse Analysis and Reflexivity', *Social Studies of Science, 16* (1986), pp. 293–307. Cf. also R. Trigg, *Understanding Social Science*, Oxford: Basil Blackwell, 1985, pp. 32–7, 86, 119, 194.

CHAPTER 1

THE PROBLEM OF ASSESSMENT

1. NEURATH AND QUINE: A PUZZLE OF HISTORIOGRAPHY

Otto Neurath's contribution to the philosophy of science and to the study of the meaning and methods of the social sciences may well appear to have been clearly defined by the end of the Seventies. For at least two decades at that time philosophical historiography had accorded Neurath the strictly positivistic position of founder and most fervent and intransigent representative of the Vienna Circle.[1] In the context of the rich and in many ways decadent culture of post-Habsburg Vienna in the Twenties and Thirties[2] he appeared as a social reformer and proponent of 'empirical enlightenment', an exuberant figure bursting with vitality, political energy and scientific optimism.

The 'physicalist' version of logical empiricism which he advanced at the beginning of the Thirties expressed, so it was thought, the most radical position of Viennese scientific enlightenment, while his programme for the unification of science and in particular his sociological epistemology appeared to unite Comtean positivism, philosophical materialism (both classic and Marxist), and Watsonian and Pavlovian behaviourism. The inflexible and in some ways ingenuous battle which he waged against metaphysics and theology was seen as obviously ideological in motivation and lacking the tools of rigorous analysis – indeed as itself being infected with metaphysics and an implicit epistemological realism.[3]

Such an understanding of Neurath's work received no small impetus from Popper's *Logik der Forschung* and more widely from the polemic successfully directed by the Popperian school against Viennese logical positivism and especially against the theory of 'protocols', sweepingly attributed to Carnap and Neurath alike.[4] Towards the end of the Seventies, however, such an interpretation was already beginning to show signs of stress. Reasons, both theoretical and historiographical, began to emerge for questioning our whole understanding of the history of mid-European logical empiricism. Today the Viennese school is coming to be seen more and more as the intertwining of various philosophical ideas and epistemological doctrines, not all of them amenable to the facile classification 'positivistic'.[5] Indeed it is coming to be thought that Neurath's own contribution, more than that of any other, admits of a non-positivistic interpretation, that it was the beginning of a rich critical development of logical empiricism which was to see its end only in 'postempiricist' epistemologies.[6]

An important indication of this had appeared in certain well-known works of Quine, principally his 'Two Dogmas of Empiricism', considered by many, alongside Gödel's theorems of incompleteness, to be the greatest contribution towards a critique of empiricism to have emerged from within the empiricist movement itself. For the traditional interpretation was faced with a serious historiographical difficulty by Quine's insistence on seeing Neurath as the impulse behind his search for an 'empiricism without dogmas'. How was it that Quine had chosen as superscription to his critique of reductionism the famous reflexive metaphor used by the 'positivist' and 'physicalist' Neurath? How, in 'Identity, Ostension and Hypostasis' and in numerous pages of *The Ways of Paradox*, *Word and Object* and *Ontological Relativity and Other Essays*, had he come to formulate an instrumentalistic, conventionalistic and 'circular' conception of science, referring directly to Neurath? According to Quine

> the philosopher's task was well compared by Neurath to that of a mariner who must rebuild his ship on the open sea. We can improve our conceptual scheme, our philosophy, bit by bit while continuing to depend on it for support; but we cannot detach ourselves from it and compare it objectively with an unconceptualized reality. Hence it is meaningless to inquire into the absolute correctness of a conceptual scheme as a mirror of reality. Our standard for appraising basic changes of conceptual scheme must be, not a realistic standard of correspondence to reality, but a pragmatic standard. Concepts are language, and the purpose of concepts and of language is efficacy in communication and in prediction.[7]

And how was it possible for the traditional interpretation to refer to a 'Duhem-Quine thesis'[8] as a pragmatic conception of science while excluding Neurath, the very writer Quine credited with being principal agent in his own encounter with Duhemian conventionalism?

How, then, are we to place Neurath today? As the most intransigent and dogmatic of the logical positivists of the Vienna Circle or, rather, as the logical positivist who succeeded in posing the very premises by which to go beyond logical positivism? How is it that an epistemological iconoclast and antipositivist *par excellence* such as Paul Feyerabend comes to write that

> in the Vienna Circle only Neurath had a clear conception of the properties of scientific research?[9]

This, at any rate, is the question which this book sets out to answer, at least to the extent that, apart from its own wider thesis, it aims to provide a contribution to the historiographical reconstruction of a number of aspects of the history of mid-European empiricism.

The attempt to answer this question will involve an overall reconsideration of Neurath's work and will re-open discussion of various commonplaces of the received interpretation, above all the significance of Neurath's intellectual association with Rudolf Carnap. It will be necessary to look anew at the relation between Neurath's 'physicalism' and the logicistic and fundamentalistic orthodoxy shared by those members of the Vienna Circle more closely linked to the teaching of Russell and Schlick. Also we shall examine the reasons for the deep epistemological and political divisions which have constantly led to the contraposition of Neurath to Popper. The current interpretation, promoted by Popper himself, finds the origin of their polemic in the ingenuousness and dogmatic close-mindedness of Neurath's positivism and Marxism and in the inconclusiveness of his *coherence theory of truth*. In other words Popper's critique of the positivistic theory of the 'empirical base' and of the programme for the unification of science which he so successfully directed against Neurath and Carnap is seen as the critical turning-point in the 'liberation' of logical empiricism. And for this the credit is held to be due to Popper himself or, at very least, to the 'supine' acquiescence of Carnap in the face of Popper's critique. It is strange to find this interpretation being swallowed even by so sharp and irreverent a critic as Imre Lakatos.[10]

2. NEURATH AND CARNAP: A MISLEADING ASSIMILATION

It has become traditional to view Carnap as having received from Neurath the vital idea of a conception of science as a 'scientific vision of the world', *wissenschaftliche Weltauffassung*. According to this view Carnap assimilated the outlines of Neurath's physicalism and encyclopedism, translating his more impassioned than rigorous style into a theoretical construct as ambitious as it was painstaking in its formation on the basis of a refined logical and linguistic analysis. Common to both was the pursuit of a 'scientific philosophy' in which the classical tradition of empiricism, from Hume to Mach, joined with the logicism of Frege and Russell to produce an intransigent polemic against the idealism and metaphysics which dominated German culture at the beginning of the century. Common to both, from the early years of the Thirties, was the physicalist ideal of a 'unified science' built on the model of post-Newtonian physics, and their consequent denial of the idealistic division between *Naturwissenschaften* and *Geisteswissenschaften.*

Both belonged, along with Hans Hahn and Edgar Zilsel, to the 'left wing' of the Vienna Circle,[11] in opposition to Schlick and other members of the Circle, such as Friedrich Waismann and Bela von Juhos, who were in more direct contact and fuller agreement with the thought of Wittgenstein.[12] Schlick, possessed of an elegant and aristocratic demeanour, lived in the style of the grand Viennese bourgeoisie, while Neurath, large and ungainly, deliberately went about unshaved in his workman's beret and had a booming voice which Schlick claimed to find intolerable.[13] Schlick, the individualist, denied the existence of 'any relationship between science and social development' and, like Wittgenstein, took not the slightest interest in political activity.[14] Neurath, on the other hand, was an active campaigner for Austrian Social Democracy, collaborated in 'Der Kampf', and in 1919 directed the Central Planning Office for the Social Democratic government of Munich. Later, in highly dramatic circumstances which were to lead to his arrest and conviction – with a sentence of eighteen months' hard labour – for collaboration in the crime of high treason, he directed the same office under the government of the Spartakist Republic of Bavaria, which came to its bloody end after only a few months of life.[15]

Carnap too termed himself a socialist and held that

> the great problems of the organisation of economy and the organisation of the world at the present time, in the era of industrialisation, cannot possibly be solved by 'the free interplay of forces', but require rational planning. For the organisation of economy this means socialism in some form; for the organisation of the world it means a gradual development towards a world government.[16]

Both were cosmopolitan anti-military pacifists. Carnap took great pains to learn Esperanto (which earned him Wittgenstein's severe reproof),[17] while Neurath invented his clever universal language of visual communication, Isotype, which was to have notable international success, especially in the United States and the Soviet Union, and which today is held to lie at the origin of modern techniques of graphic representation of statistical data.[18]

This list of similarities between the two on which the traditional interpretation is built could well be prolonged. But the importance of such similarities lies above all in the personal and psychological element which they reveal: Carnap, the 'introverted and cerebrotonic' intellectual, in Hempel's phrase, felt the charm of Neurath's abounding intellectual and political drive, admired the richness and variety of his interests and of his extensive reading, felt an emotional pull to his socialist positions, and even showed himself disposed, in the light of Neurath's criticisms, to reformulate some aspects of his original (and over-logicistic and academic) plan for a 'logical construction of the world'.[19] On one essential point, however, Carnap maintained a theoretical stance which was in profound disagreement with that of Neurath, as is brought into sharp relief by certain passages of his intellectual autobiography:

> For Neurath the aim of a unified science was of vital importance. The sharp distinction between natural sciences and *Geisteswissenschaften* (humanities), which was strongly emphasized in contemporary German philosophy, was in his view an obstacle on the road towards our social goal. [...] He would deride those purist philosophers who sit on their icy glaciers and are afraid they might dirty their hands if they were to come down and tackle the practical problems of the world.
>
> In spite of the difference of opinion between Neurath and the other members of the Circle at certain points, we certainly owed very much to his collaboration. Of particular importance for me personally was his emphasis on the connection between our philosophical activity and the great historical processes going on in the world: Philosophy leads to an improvement in scientific ways of thinking and thereby to a better understanding of all that is going in the world, both in nature and in society; this understanding in turn serves to improve human life. In numerous private conversations I came into even closer contact with Neurath's ideas. He liked to spice the talks with a lot of wit and sarcasm, criticizing the views and attitudes of others, including myself and of philosophers whom I appreciated highly, such as Schlick and Russell. These talks were always most lively and stimulating; and in spite of my frequent opposition, I learned a great deal from them. [...]
>
> One of the important contributions made by Neurath consisted in his frequent remarks on the social and historical conditions for the development of philosophical conceptions. He criticized strongly the customary view, held among others by Schlick and by Russell, that a wide-spread acceptance of a philosophical doctrine depends chiefly on its truth. He emphasized that the sociological situation in a given culture and in a given historical period is favorable to certain kinds of ideology or philosophical attitude and unfavorable to others.[20]

Up to this point Neurath's opinions met with no great disagreement since all the members of the Circle, Carnap says, were profoundly interested in social and political progress and, for the most part, termed themselves 'socialists'. But Neurath went further: to support the desirability or otherwise of certain logical or empirical propositions, he was in the habit of drawing arguments more of a pragmatic or political than of a theoretical nature.[21] On this point disagreement between Carnap and Neurath was fundamental and it led to an increasing tension between them. Their correspondence offers dramatic testimony to this.[22] The disagreement was not, and could not be, overcome, because, as Carnap again says:

> we liked to keep our philosophical work separated from our political aims. In our view, logic, including applied logic, and the theory of knowledge, the analysis of language, and the methodology of science, are, *like science itself, neutral* with respect to practical aims, whether they are moral aims for the individual, or political aims for a society. Neurath criticized strongly this *neutralist attitude*, which in his opinion gave aid and comfort to the enemies of social progress. We in turn insisted that the intrusion of practical and especially of political points of view would *violate the purity of philosophical methods*.[23]

The close historiographical linking of the work of Neurath and Carnap has tended to overshadow this crucial point of theory which divides Neurath from almost all the other members of the Viennese group and has succeeded in obscuring the specific characteristics of Neurath's epistemological contributions. Significant consequences of this will become apparent in the course of this study.

A further reason for the emergence of this misappraisal lies in the fact that, in the direct comparison of Neurath with Carnap, only Carnap has been seen as the rigorous thinker, thoroughly grounded in logic and linguistics,[24] while Neurath has received recognition almost only for his qualities of intellectual instigator, clever 'propagandist' and cultural entrepeneur.[25] Furthermore, in contrast to Carnap and the great majority of the Circle, Neurath concerned himself less extensively with logic and philosophy of science in the strict sense than with sociology and the epistemology of the social sciences (in addition to economics, militant politics and techniques of visual communication), none of them subjects, it may be said, to which students of logical neopositivism have accorded any more than scant regard.

3. NEURATH AND POPPER: AN EPISTEMOLOGICAL AND POLITICAL POLARITY

Notwithstanding the warm appreciation of Neurath's early sociological studies shown by Max Weber in his *Wirtschaft und Gesellschaft*,[26] it remains that Neurath's reputation as a sociologist and as a student of the social sciences underwent something of an eclipse, even in the Forties and Fifties when empiricism was at the peak of its influence on the philosophy of the social sciences in Europe and the United States. The only accurate and systematic – albeit in large part severely critical – study of Neurath's sociological thought appeared in 1967, the work of an Italian writer, Gianni Statera.[27] Statera concentrated, however, on the aspects of Neurath's work held in common with the positivist epistemology of the Vienna Circle, underestimating the very real differences between them and almost completely ignoring the polemic of Neurath's confrontations with Popperian falsificationism.

Nor has the position been markedly different within the tradition of Western Marxism. This is a school to which Neurath may be said in many ways, politically no less than intellectually, to belong – when viewed as the Austrian social democrat, the economist and student (in company with Karl Korsch, Anton Pannekoek and Otto Bauer) of the problems of 'socialization', and as the philosopher of Austro-Marxism in an epistemological tradition akin to that of Friedrich and Max Adler. And yet it would not be too much to say that Neurath has received from this school complete neglect.

Neurath's philosophical thought may well also have seemed consigned to oblivion after the 'revolt against positivism',[28] in England and America in the Sixties, which set off a widespread crisis in the 'standard view' of empiricism and in its extension into the field of social sciences. This has in fact been the fate of several contemporary (or near-contemporary) 'positivist' methodologists, psychologists, and sociologists, such as E. Zilsel, G. Bergmann, the American G.A. Lundberg, and Herbert Feigl himself. But it is, on the contrary, this very crisis in Anglo-American empiricism, along with the rise since the second half of the Sixties of a historical approach to the traditional problems of the philosophy of science, which now seems to assert the present-day relevance of many aspects of Neurath's thought. And these are aspects which will not only be sought in vain in the work of the major exponents of mid-European logical empiricism but are often views which Neurath expressed in direct conflict with many of those exponents, with Carnap, for example, or Schlick, or the early Wittgenstein, with Russell, Reichenbach, or with Tarski.

The historiographical thesis which will be advanced in the present work is therefore the following: that Neurath's epistemological thought stands so considerably at variance with what has historically come to be understood as 'logical empiricism'[29] that his contribution has an importance to present-day philosophy of science which is directly proportionate to the lack of attention it has in fact so far received. Neurath's writings are today, after Hanson, Lakatos, Bohm, Kuhn, Mary Hesse, and Feyerabend, and even after Wittgenstein in his late phase, surprisingly rich in lessons for those who have taken account of the irreversible crisis in 'dogmatic' empiricism, but are still little inclined to accept that the collapse of the foundations of scientific knowledge can be taken as a basis for, and an opportunistic justification of, every sort of 'knowledge', including even the most irrational, dogmatic, or speculative.

Neurath's teaching is valuable notwithstanding – indeed to a large extent because of – the meagre propensity he had for minute and systematic analysis which has so often been held against him. If, as has been said, one of the main reasons for the rapid exhaustion and break-up of the Vienna Circle was its focussing on *minutiae*, the waning of its interest in the general problems of knowledge, science and society, in favour of a scholasticism applied to logical and linguistic puzzles – and here it is difficult for Carnap and his school not to spring immediately to mind – then it is plain that here too there is yet another reason why Neurath should not figure in the disintegrative crisis of logical positivism and of neo-empiricism.[30]

But all this could still seem insufficient to justify a resumption of philosophical or theoretical, as opposed to merely historiographical, interest in the work of Neurath. The theoretical thesis of the present work is therefore as follows: that the fundamental contribution of Neurath's epistemological thought lies in his critique, empiricist no less than conventionalistic, of 'pseudorationalism'. This critique, born of the teaching of Pierre Duhem, Henri Poincaré, and Abel Rey, of Kazimierz Ajdukiewicz's pragmatic notion of language, and of non-deterministic philosophical approaches to quantum physics, makes of Neurath's epistemology and sociology something which transcends not only classical neopositivist *Erkenntnislehre* but also the epistemology of Popper and of the various Popperian and neo-Popperian schools of today. Neurath's philosophy lays the foundations of a thesis opposed not only to Popper's *Logik der Forschung* but also to his famous sociological and political ideas elaborated in *The Poverty of Historicism* and *The Open Society and Its Enemies*, ideas inspired by the individualism of Friedrich von Hayek and severely critical not just of Marxism but also of any holistic approach to the theme of social change.

In the light of these epistemological and political aspects Neurath will, it is hoped, re-emerge, after years of unjustified neglect, as one of the fullest, most complete and creative characters of Viennese culture of the first half of the century – a thinker in many ways opposed to Wittgenstein, but, notwithstanding the apparent linearity and worldly aspect of his intellectual development, no less profound, aware, or original than him and certainly no less dramatic.

No better witness to the value and significance of Neurath's work could be expected, after years of fierce philosophical polemic and of deep political hostility, than Popper himself. There appeared in 1973 a late, but not for that reason any less important, testimony by him in his contribution to 'Memories of Otto Neurath',[31] in which he describes Neurath as:

> a big, tall, exuberant man with flashing eyes, a big red beard, and a loud voice. The impression was of a most unusual personality, of a man of tremendous vitality and drive, of a man who believed passionately in his social, political and philosophical theories, but who believed even more in himself, though very ready to laugh about himself; of a man who was immensely attractive, but who cared nothing about it; who would not look behind him or, when rushing ahead, care very much about whom his big stride might knock down.[32]

Popper sees in Neurath the true founder of the Vienna Circle and the architect of its success:

> I have little doubt that it was Otto Neurath who, with the hope in mind of a philosophical reform of politics, attempted to give to the circle of men around Schlick and Hahn a more definite shape: and thus it may have been he, perhaps more than anybody else, who was instrumental in turning it into the 'Vienna Circle'. [...] he enjoyed his role in the Circle, in which he was the strongest personality, even though he was not its philosophical leader.
>
> It was largely due to Neurath that the Circle became famous. Neurath inspired the founding of *Erkenntnis*, edited by Carnap and Reichenbach; he worked for co-operation with similar groups on an international scale: he organised its congresses, got a plan for regular 'Congresses for Scientific Philosophy' accepted; and founded the 'Unity of Science Movement' and the 'International Encyclopedia for Unified Science'.[33]

He concludes:

> I hoped that the new philosophy of science would be able to make important contributions to political theory, but, in the main, contributions critical of those very ideas which Neurath still adhered to (such as historical prophecy, and many others). So whenever we met, there was a head-on collision in precisely that field which interested him most; and this made him suspicious of my philosophical ideas. [...] Neurath and I disagreed deeply on many and important matters which interested us both except one: the view that the theory of knowledge was important for an understanding of history and of political problems. Yet though we had disagreed so deeply about so many and so important matters, I shall always feel that he was one of the strongest personalities I ever met; a real original thinker, and an undaunted fighter who dreamt of a better and more humane world.[34]

Here, then, we see Neurath viewed not as the pale reflection of Carnap nor as the Vienna Circle's 'man of action', but as the most original thinker and the strongest personality in Viennese intellectual society – an adversary fully worthy of philosophical respect and esteem.

NOTES

[1] For Neurath's biography see *ES, passim*. See also P. Neurath, 'Otto Neurath und die Soziologie', in R. Haller (ed.), *Schlick und Neurath. Ein Symposion, Grazer Philosophische Studien*, 1984, Amsterdam: Rodopi, no. 16–17, pp. 223–40.
[2] Cf. A. Janik, S. Toulmin, *Wittgenstein's Vienna*, London: Weidenfeld and Nicolson, 1973. For references to Neurath see pp. 133 ff.
[3] Among the numerous upholders of this interpretation see C.G. Hempel, 'Logical Positivism and the Social Sciences', in P. Achinstein and S.F. Barker (eds.), *The Legacy of Logical Positivism*, Baltimore: The Johns Hopkins Press, 1969, pp. 163–94; A.J. Ayer, *Philosophy in the Twentieth Century*, London: Weidenfeld and Nicolson, 1982, pp. 121–30, 138–41; F. Barone, *Il positivismo logico*, Roma and Bari: Laterza, 1977, pp. 307–18; J.W.N. Watkins, 'Otto Neurath', *British Journal for Philosophy of Science, 25* (1974), pp. 343–52. But see also, for Hempel and Barone's recent reconsideration of Neurath's philosophy of science, C.G. Hempel, 'Schlick und Neurath: Fundierung versus Kohrenz in der wissenschaftlichen Erkenntnis', in R. Haller (ed.), *Schlick und Neurath. Ein Symposion*, pp. 1–18; F. Barone, 'Protocol Sentences and Scientific Anarchism', ibid., pp. 327–45.
[4] K.R. Popper, *Logik der Forschung*, Wien: Julius Springer, 1935, pp. 53–4, English trans. *The Logic of Scientific Discovery*, London: Hutchinson, 1968, pp. 95–6.
[5] Cf.: H. Berghel, A. Hübner and E. Köhler (eds.), *Wittgenstein, der Wiener Kreis und der Kritische Rationalismus*, Akten des Dritten Internationalen Wittgenstein Symposiums 1978, Wien: Hölder-Pichler-Tempsky, 1979; G.H. von Wright, *Wittgenstein*, Oxford: Basil Blackwell, 1982; *Le Cercle de Vienne: doctrine et controverses*, Journées Internationales, Paris 29–30 septembre 1983, proceedings forthcoming in *Fundamenta Scientiae*, Pergamon Press; R. Haller, 'New Light on the Vienna Circle', *The Monist, 65* (1982), no. 1, pp. 25–35.
[6] Cf.: H. Haller, 'Das Neurath-Prinzip. Grundlagen und Folgerungen', in F. Stadler (ed.), *Arbeiterbildung in der Zwischenkriegszeit*, Wien and München: Löcker Verlag, 1982, pp.79–87; H. Rutte, 'Der Philosoph Otto Neurath', ibid., pp. 70–8; R. Hegselmann, 'Otto Neurath, Empiristischer Aufklärer und Sozialreformer', in O. Neurath, *Wissenschaftliche Weltauffassung, Sozialismus und Logischer Empirismus*, ed. by R. Hegselmann, Frankfurt a.M.: Suhrkamp, 1979; F. Stadler, *Vom Positivismus zur 'Wissenschaftlichen Weltauffassung'*, Wien and München: Löcker Verlag, 1982; F. Stadler, 'Otto Neurath – Moritz Schlick: Zum philosophischen und weltanschaulich-politischen Antagonismus im Wiener Kreis', in *Schlick und Neurath. Ein Symposion*, pp. 451–64; H. Rutte, 'Über Neuraths Empirismus und seine Kritik am Empirismus', ibid., pp. 365–84; R. Haller, 'Zwei Arten der Erfahrungsbegründung', ibid., pp. 19–33; G. Giorello, 'Il falsificazionismo di Popper', in *Storia del pensiero filosofico e scientifico*, ed. by L. Geymonat, vol. VII, Milano: Garzanti, 1976, pp. 456–518; R. Haller, 'Il primo Circolo di Vienna', in A. Gargani (ed.), *Il Circolo di Vienna*, Ravenna: Longo Editore, 1964, pp. 45–61.

[7] W.V.O. Quine, 'Identity, Ostension, and Hypostasis', *Journal of Philosophy*, 1950, now also in W.V.O. Quine, *From a Logical Point of View*, Cambridge (Mass): Harvard University Press, 1980, p. 79. Cf.: W.V.O. Quine, *The Ways of Paradox and Other Essays*, New York: Random House, 1966, pp. 210, 212–3; W.V.O. Quine, *Word and Object*, Cambridge (Mass): Technology Press of MIT and John Wiley, 1960, pp. 3, 123–4; W.V.O. Quine, *Ontological Relativity and Other Essays*, pp. 16, 84, 127.

[8] On the 'Duhem-Quine thesis' see for instance: I. Lakatos, 'Falsification and the Methodology of Scientific Research Programmes', in I. Lakatos and A. Musgrave (eds.), *Criticism and the Growth of Knowledge*, pp. 184–6; M. Hesse, 'Quine, Duhem and a New Empiricism', in S.G. Harding (ed.), *Can Theories Be Refuted? Essays on the Duhem-Quine Thesis*, Dordrecht and Boston: D. Reidel, 1976.

[9] Cf.: P.K. Feyerabend, *Problems of Empiricism*, Cambridge: Cambridge University Press, 1981, vol. 2, p. 86; P.K. Feyerabend, 'Against Method: Outline of an Anarchist Theory of Knowledge', *Minnesota Studies for the Philosophy of Science*, ed. by H. Feigl and G. Maxwell, *4* (1970), pp. 71–3 ("[Today] it is also admitted, in complete agreement with Neurath's original views, that *both* theories *and* observations can be abandoned: theories may be removed because of conflicting observations, observations may be removed for theoretical reasons"). H. Feigl himself, who may be considered the empiricist philosopher most faithful to the Vienna Circle's original theses, recognises that: "After decades of criticism, these 'sense-data' and 'pointer-readings' doctrines have been largely abandoned. Some philosophers of science again flirt with Otto Neurath's suggestion that logical reconstruction – from scratch – (i.e. from an ultimate and indubitable basis) is chimerical. Neurath himself, in order to illustrate the lines of scientific progress, used the analogy of rebuilding a ship on the high seas" (H. Feigl, 'Empiricism at Bay?', *Boston Studies in the Philosophy of Science*, ed. by R.S. Cohen and M.W. Wartofsky, 14, 1971, now also in H. Feigl, *Inquiries and Provocations: Selected Writings 1929–1974*, ed. by R.S. Cohen, Dordrecht and Boston: D. Reidel, 1980, pp. 273–4).

[10] I. Lakatos, 'Falsification and the Methodology of Scientific Research Programmes', in I. Lakatos and A. Musgrave (eds.), *Criticism and the Growth of Knowledge*, pp. 113–4, 156.

[11] Cf. H. Neider, 'Memories of Otto Neurath', *ES*, pp. 45–9, 82, note 20. In this last footnote the editors of *ES* tell us that "Neurath made frequent interjections, 'metaphysics!', during the Circle's reading and discussion of Wittgenstein's *Tractatus*, to the irritation of Moritz Schlick who finally told him he was interrupting the proceedings too much. Hans Hahn, as conciliator, suggested to Neurath just to say 'M' instead. After much humming – so C.G. Hempel was later told – Neurath made another suggestion to Schlick: 'I think it will save time and trouble if I say 'non-M' every time the group is *not* talking metaphysics'". See also R. Haller and H. Rutte, 'Gespräch mit Heinrich Neider, Wien: Persönliche Erinnerungen an den Wiener Kreis', in J.C. Marek, J. Zelger, H. Ganthaler and R. Born (eds.), *Österreichische Philosophen und ihr Einfluß auf die analytische Philosophie der Gegenwart*, Sonderband *Conceptus*, no. 28–30, Innsbruck 1977.

[12] Cf.: E. Nemeth, *Otto Neurath und der Wiener Kreis*, Frankfurt a.M. and New York: Campus Verlag, 1981, pp. 55–60. See also: F. Stadler, 'Aspekte des gesellschaftlichen Hintergrunds und Standorts des Wiener Kreises am Beispiel der

Universität Wien', in *Wittgenstein, der Wiener Kreis und der Kritische Rationalismus*, pp. 53–9; R. Hegselmann, *Normativität und Rationalität. Zum Problem praktischer Vernunft in der analytischen Philosophie*, Frankfurt a.M. and New York: Campus Verlag, 1979, pp. 29–41; R. Haller, 'Der 'Wiener Kreis' und die analytische Philosophie', in R. Haller, *Studien zur österreichischen Philosophie*, Amsterdam: Rodopi, 1979; W.M. Johnston, *Österreichische Kultur- und Geistesgeschichte. Gesellschaft und Ideen im Donauraum 1848–1938*, Wien: Böllau, 1978; E. Mohn, *Der logische Positivismus. Theorien und politische Praxis seiner Vertreter*, Frankfurt a.M: Campus Verlag, 1977.

[13] Cf. H. Neider, 'Memories of Otto Neurath', *ES*, pp. 47–8.

[14] Ibid., p. 47.

[15] See on this subject the testimony of W. Schumann, in 'Memories of Otto Neurath', *ES*, pp. 15–7. Schumann reports that, in order to accept the political task offered to him by the Social Democratic Party, Neurath had to make the decision to abandon his academic career. At that time Neurath was associated with the sociology department of the University of Heidelberg, whose chairman was Max Weber. Later Weber took part in the political trial which followed the repression of the Soviet Bavarian Republic, by making a testimony partially in favour of Neurath. Otto Bauer, who at that time was the foreign Minister of Austria, also made an intervention in his favour. See in *ES*, pp. 18–28, the pages of Neurath's autobiography concerning his experience of the economic socialisation of Bavaria. For further details see the accurate historical reconstruction of Neurath's participation in the activities of the Soviet Bavarian Republic in R. Hegselmann, 'Otto Neurath. Empiristischer Aufklärer und Sozialreformer', pp. 23–33. See in addition: the historical documentation gathered by G. Schmolze (ed.), *Revolution und Räterepublik in München 1918/19 in Augenzeugenberichten*, Düsseldorf: Rauch, 1969; R.M. Lepsius, 'Max Weber in München', *Zeitschrift für Soziologie, 6* (1977), pp. 103–118; K. Fleck, *Otto Neurath. Eine biographische und systematische Untersuchung*, doctoral dissertation in the Geisteswissenschaftliche Fakultät of Graz University, Graz 1979.

[16] R. Carnap, 'Intellectual Autobiography', in P.A. Schilpp (ed.), *The Philosophy of Rudolf Carnap*, La Salle (Ill): Open Court, 1963, p. 83.

[17] Ibid., p. 26.

[18] Cf.: O. Neurath, 'From Vienna Method to Isotype', unfinished manuscript, 1945, *ES*, pp. 214–48; O. Neurath, *Modern Man in the Making*, New York: A.A. Knopf, 1939. See also: R. Hegselmann, 'Otto Neurath. Empiristischer Aufklärer und Sozialreformer', pp. 47–52; R. Kinross, 'Graphic Communication and the Work of Otto Neurath', unpublished paper, Journées Internationales sur *Le Cercle de Vienne: doctrine et controverses*, Paris 29–30 September 1983.

[19] C.G. Hempel, 'Logical Positivism and the Social Sciences', p. 165. According to Hempel "Neurath was an activist and planner of tremendous energy; [...] Neurath was extroverted, ebullient, and extremely vivacious. He was a heavy but very dynamic and most engaging man, who often signed letters to his friends with a drawing of a cheerful elephant with Neurath's initials branded on his hindquarter, holding a bunch of flowers for the addressee in his trunk" (ibid.). Philipp Frank reminds us that "Otto Neurath even enrolled for one year in the Divinity School of the University in order to get an adequate picture of Catholic philosophy, and won an award for the best paper on moral theology" (*Modern Science and Its Philosophy*, p.2).

[20] R. Carnap, 'Intellectual Autobiography', pp. 23, 24, 22.

[21] For similar evaluations cf. C.G. Hempel, 'Logical Positivism and the Social Sciences', pp. 166–7, 173, 175.

[22] The expected publication of the Neurath-Carnap correspondence (ed. by H.L. Mulder and M. Neurath, Dordrecht and Boston: D. Reidel) will shed light on this aspect. However the numerous and important letters included in the *Carnap Collection* (Hillman Library, University of Pittsburgh, Pennsylvania) show the increasing theoretical disagreement between Carnap and Neurath. In the last years, particularly in 1943–45, it burst into a sharp and bitter polemic between them. Neurath openly criticised Carnap's theory of confirmation, induction and semantics as a kind of logico-linguistic scholasticism and realistic (Aristotelian) metaphysics. Carnap strongly opposed the publication of Neurath's *Foundations of the Social Sciences* in the second volume of the *International Encyclopedia of Unified Science*.

[23] R. Carnap, 'Intellectual Autobiography', p. 23 (italics added).

[24] Cf.: C.G. Hempel, 'Logical Positivism and the Social Sciences', pp. 174–5; C.G. Hempel, 'Rudolf Carnap, Logical Empiricist', *Synthese*, *25* (1973), pp. 261–2. Hempel describes Neurath as "an influential figure in the earlier stages of logical empiricism, whose formulations were suggestive but vague; whose argumentation, though provocative and often persuasive, tended to be loose, sketchy, and programmatic". See also: C.G. Hempel, 'Der Wiener Kreis: eine persönliche Perspektive', in H. Berghel, A. Hübner and E. Köhler (eds.), *Wittgenstein, der Wiener Kreis und der Kritische Rationalismus*, pp. 21–6; J.W.N. Watkins, 'Otto Neurath', p. 345.

[25] H. Feigl. 'The *Wiener Kreis* in America', in D. Fleming and B. Bailyn (eds.), *The Intellectual Migration 1930–1960*, Cambridge (Mass): Harvard University Press, 1969, now also in H. Feigl, *Inquiries and Provocations: Selected Writings 1929–1974*, p. 62.

[26] M. Weber, *Wirtschaft und Gesellschaft*, Tübingen: J.C.B. Mohr, 1956, English trans. *Economy and Society*, Berkeley: University of California Press, 1978, pp. 104–5, 106–7, 111, 207–8.

[27] G. Statera, *Logica, linguaggio e sociologia. Studio su Otto Neurath e il neopositivismo*, Torino: Taylor, 1967.

[28] Cf. D. Shapere, 'Meaning and Scientific Change', in R.G. Colodny (ed.), *Mind and Cosmos*, Pittsburgh: University of Pittsburg Press, 1966, pp. 41–65.

[29] For a presentation of the neopositivistic commonsense background see H. Feigl's following essays: 'Logical Empiricism', in D.D. Runes (ed.), *Twentieth Century Philosophy*, New York: Philosophical Library, 1943, now also in H. Feigl and W. Sellars (eds.), *Readings in Philosophical Analysis*, New York: Appleton-Century-Crofts, Inc., 1949; 'The Origin and Spirit of Logical Positivism', in P. Achinstein and S.F. Barker (eds.), *The Legacy of Logical Positivism*, pp. 3–24; 'The *Wiener Kreis* in America'; 'Empiricism at Bay?'.

[30] Cf.: J. Passmore, 'Logical Positivism', in *Encyclopedia of Philosophy*, ed. by P. Edwards, vol. 5, London: Macmillan, 1967, p. 56; K.R. Popper, 'Autobiography', in P.A. Schilpp (ed.), *The Philosophy of Karl Popper*, p. 71.

[31] *ES*, pp. 1–83.

[32] Ibid., p. 52.

[33] Ibid., pp. 54–5.

[34] Ibid., pp. 55–6.

CHAPTER 2

ENLIGHTENMENT, NEO-MARXISM, CONVENTIONALISM: TOWARDS A CRITIQUE OF CARTESIAN RATIONALISM

1. SCIENCE AS A 'MEANS FOR LIFE'

The composition of the Vienna Circle's 'manifesto' – *Wissenschaftliche Weltauffassung. Der Wiener Kreis* – posed a threat, as indeed Neurath intended it should, to the philosophical hegemony of Schlick. The circumstances which made it a shrewd political move against him are well known.[1] There has also been much insistence on the temperamental differences which made even personal relations difficult between Neurath and Schlick,[2] differences somewhat similar to those which hindered the great majority of the Vienna Circle from developing relations with Wittgenstein.[3] The attention given to the anecdotal aspect of these problems of personality has succeeded, however, in blurring the specific philosophical differences which separated Neurath from all the other members of the group, even in matters of editorial or cultural policy where Neurath aimed to present to the outside world an image of unity within the movement.[4]

Most characteristic of Neurath, separating him forcefully even from Carnap (who was also party to the 'manifesto' of '29), was his strong concern for the 'questions of social life' (*Lebensfragen*) which, though they did not lie at the centre of the Circle's theoretical research, he nevertheless saw as the implicit goal and motivating force behind epistemological enquiry. Among these he included, for instance, the question of the reorganisation of the economy and society, the reform of education and schooling, and the unification of mankind.[5]

From the point of view of Neurath's strategy, the significance of *Wissenschaftliche Weltauffassung* lay solely in its force as *novum organon* of a 'rational transformation of the social and economic order' (*eine rationale Umgestaltung der Gesellschafts- und Wirtschaftsordnung*).[6] For him the assault on metaphysics and theology or the goal of the 'unification of science' as an outcome of collective theoretical effort, or the logical and linguistic clarification of traditional problems of philosophy and their partial transformation into empirical problems were all aspects of a new *Aufklärung*[7] and a new encyclopedism motivated at base by political and social rather than theoretical considerations.

This accorded well with Neurath's view of modern empiricism as the direct heir of materialism. He saw it as a philosophy rooted in the real world, intended to shape the forms of public and private life and even to be the guiding force behind social and economic struggles carried out in Europe in the name of socialism by 'large popular masses'.[8] And it is significant that he re-asserted in no uncertain terms the centrality of these notions at the very end of the 1929 'manifesto', a text which remained faithful, however, in its individual epistemological themes, to the substance of Viennese orthodoxy and its positivistic antecedents, i.e. Russell's logical fundamentalism, Wittgenstein's claim to resolve all philosophical problems through logical and linguistic analysis, Schlick and Carnap's notion of the 'meaning' of scientific statements lying in the possibility of their being reduced to sentences on 'immediate data' (*das Gegebene*), and the attempt to make this the basis of a rigorous demarcation between science and metaphysics.[9]

In open disagreement with Schlick and his more enthusiastic disciples, Neurath argued that the Vienna Circle's theoretical work lay 'close to the problems of contemporary life' (*dem Leben der Gegenwart nahe*).[10] Their activities were, he said, interwoven with the political struggles and social aspirations of the day and, like these, would meet with strong opposition. And this remained true despite the fact that not all the members of the Circle had a natural vocation to be 'combatents' (*Kämpfer*) and that some even exhibited a preference for solitary exclusion on the 'icy slopes of logic' (*auf den eisigen Firnen der Logik*), disdaining to 'intermingle with the masses' and deploring the 'vulgarisation' (*Trivialisierung*) inevitably entailed by the popularisation of philosophical questions.[11]

It is important to stress that in this general orientation of Neurath's thought, strongly weighted as it was from an evaluative point of view, enlightenment and neo-Marxism (i.e. a 'scientific' and sociological Marxism)[12] were taken to advocate an instrumentalistic conception of scientific knowledge and at the same time to propose a strong historical and sociological relativisation of the methods and achievements of science. We are dealing, then, with a position which could hardly fail to come swiftly into conflict with established Viennese logicistic and positivistic orthodoxy in that it exposed Neurath's thought – in contrast to that of the great majority of the members of the Circle, with the only sure exception of Philipp Frank – to the direct and deep influence of classical conventionalism, especially that of Pierre Duhem, Henri Poincaré, and Abel Rey.[13] This influence was to end by cancelling out not only the influence of logicism, but also that of Comtean positivism.

The criticism which, in a three-fold succession of increasing philosophical importance, Neurath directed over the first half of the Thirties against, in turn, Schlick, Carnap, and then Popper is confirmation in itself of the extent to which the instrumentalistic and conventionalistic elements in his epistemological thought distanced him from the fundamentalist and justificationist *Erkenntnislehre* of the Vienna Circle's orthodoxy.

2. SCIENTIFIC HOLISM

Instrumentalistic and conventionalistic elements were, however, already to be seen in works on philosophy and methodology which Neurath published prior to the formation of the Vienna Circle. Also present at that time was the element which was to became central to his thought in the Thirties – the critique of 'pseudorationalism'.

In 1910, in his publication entitled 'Zur Theorie der Sozial-wissenschaften', a careful examination of *Logik* by Wilhelm Wundt,[14] we find Neurath stating for the first time his idea of the unity of science which in the Thirties was to become the most characteristic and successful of the Circle's theoretical proposals. Even in this early work we find Neurath combining the influence of Comte, from whom he derived the idea itself of the unitary nature of scientific knowledge, with conventionalistic elements drawn from his reading of the major works of Duhem and Poincaré.[15]

While re-affirming the claims of Comte's encyclopedic conception, Neurath nevertheless explicitly distanced himself from its more specific philosophical implications.[16] He saw the unity of science not as the accumulation of specialist researches (*Sammlung von Spezialarbeiten*) nor, as Comte had proposed, as the hierarchical organisation of these researches on the basis of philosophical premises. His own view was that unity should be founded on the common principles (*gemeinsame Prinzipien*) which actually governed scientific work. Far from reiterating Comte's professed desire to take science out of the hands of the scientists and to give it over to the philosophers, Neurath envisaged an enquiry emanating from the very heart of scientific activity, one which would be the work of scientists capable of so pushing themselves beyond the constraints of their discipline as to be able to comprehend the whole panorama of science.[17]

He saw general theoretical work by such scientists as Jevons, Pearson, Enriques, or Wundt himself to be a first step in this direction.[18] But to follow such initiatives up required the collective work of a plurality of scientists

(*Zusammenarbeiten mehrerer Gelehrter*), capable not only of pooling their specialist abilities or of forming reciprocal links between different theories but also of grasping in its immediacy the comprehensive system of the theoretical relationships of science (*ganzes System von Relationen*).[19]

It is particularly interesting to see Neurath making evident use, in this early formulation of 'unified science', of a suggestion he drew from Duhem. This was the 'holistic' conception of science as consisting of propositions which admit of being tested only in respect of their totality and not in respect of each individual proposition alone.[20] Moreover, a conventionalistic conception of this totality was also implied: the job of scientists, he observed, was to decide from case to case whether the comprehensive theoretical system (*Gesamtsystem der Theorie*) was to be preserved, resorting to supplementary hypotheses (*Zusatzhypothesen*) to account for otherwise unaccountable phenomena, or whether it was more appropriate (*zweckmäßiger*) for the entire scientific system to be built afresh on new bases (*das ganze System neu zu gestalten*).[21]

It should also be noted that, compared with Comte's original outline, Neurath's concerns were rather more pragmatically orientated. Like Saint-Simon and Comte, Neurath understood science to express the need for an all-embracing vision of the world,[22] but, unlike them, he derived from this not so much a wholesale philosophy of science as a pragmatic aim: his prime objective was to achieve 'an organisation of scientific work' (*eine Organization der wissenschaftlichen Arbeit*) which would overcome the 'chaos' caused by the disorganised profusion of specialisations and the 'chaotic' dispersion of scientific work brought about by the isolation of the individual researcher.[23]

To Neurath's mind the theoretical unification of science – attainable solely by means of a practical collaboration betweeen scientists – was itself the means by which a general cooperation based on a rational and productive division of labour (*gedeihliche Arbeitsteilung*) would be possible in the field of scientific research. What concerned him most was the rational organisation and productivity of scientific work and the goal of ensuring that each researcher should consciously aim to contribute to the most important scientific breakthroughs.[24]

3. A CONVENTIONALISTIC CRITIQUE OF CARTESIAN 'PSEUDORATIONALISM'

Three years later, at an important stage in the development of his epistemological thought, Neurath adopted in his 'Die Verirrten des Cartesius und das Auxiliarmotiv' a clear anti-fundamentalist stance in opposition to what he termed Descartes' 'pseudorationalism'.[25]

Descartes' methodology, he argued, started from an entirely unacceptable distinction between the spheres of theoretical research and practical action. With regard to practical and moral action, Descartes, employing his famous metaphor of travellers lost in a wood,[26] had granted that it was very often necessary to work without being able to take account of the entire scope of available alternatives, that it was necessary to act on the basis of partial knowledge or of provisional rules without waiting for evidence or certainty. With regard to theoretical enquiry, on the other hand, he had maintained the opposite, that it was possible, by adopting a method made appropriate and 'justified' on metaphysical grounds, to achieve a definitive understanding of truth and to provide firm ground for human knowledge.

In opposition to this rigid contradistinction between the spheres of action and cognition, Neurath maintained that uncertainty of cognitive premises and precariousness of results was common to both areas of experience and that at most a difference of degree could be admitted between them, but not a difference of principle.[27] Even theoretical thought, he argued, operated in many different forms on the basis of 'provisional rules'.[28]

Above all, he maintained, it was necessary to take account of practical limitations: the brevity of human life, for example, forces us to define our philosophical positions within the restrictions of time, and these can hinder the elaboration of even a single hypothesis should it require any complex process of development and testing.[29] But it was in the end reasons of principle which led to the belief that a *Weltanschauung* or scientific system could be constructed "only by one who operates on the basis of uncertain premises".[30] Any attempt to establish a general philosophy (*Weltbild*) starting from a *tabula rasa* and seeking to join new propositions to those initially held to be definitively true could in reality only be full of illegitimate assumptions (*Erschleichungen*). Descartes' ideal of clarity and distinction, together with his claim to take mathematical demonstration as a general methodological rule of knowledge had therefore to be false. The phenomena which science attempts to explain were interconnected in such a complex way that it was impossible to try to describe them by means of a kind of 'unidimensional chain of propositions' (*durch eine unidimensionale Kette von Sätzen*). Rigorous 'distinction' between concepts was an unattainable goal, for the reason that every statement contains implicit references to an infinite number of other statements – here again we see Neurath's holistic approach in operation – and depends for its validity on the validity of all the others. Every concept intertwines itself with all the conceptual sequences which precede it, and all attempts to introduce fundamental change into our conceptual universe are themselves the circular product of our preceding conceptualisations.[31]

Furthermore, the provisional nature of conceptual categories was closely bound in with the plurality of theoretical hypotheses which always accompanies the attempt to explain or predict an empirical event. Only the judgment of the researcher, and not the application of some rigorous method, could bring about a reduction in the number of equally probable theoretical hypotheses, as proved to be the case, for example, with the theory of light.[32] Besides, thought was an indivisible 'psychological unity' (*eine psychologische Einheit*) within the development of each individual, and it was only in a highly restricted sense that we were able to refer to intellectual processes in isolation. Descartes, however, had treated development of thought as if he had been dealing with a system of logical relationships (*ein System logischer Beziehungen*) entirely unaffected by psychological conditioning and as if it were always possible in philosophical research to begin again from scratch, starting from totally uninfluenced positions.[33]

For all of these reasons Neurath argued that the model of rationality proposed by Descartes had to be labelled as 'pseudorationalistic'. Cartesian 'pseudorationalism', in his submission, started from the fallacious assumption that for every philosophical question there existed *one and only one* rational solution and that this solution could in all circumstances be discovered by means of a methodologically correct inquiry. 'Pseudorationalism' therefore did rationalism a bad turn, in as much as its claim to certainty and univocality led in fact to self-deception and pretence. 'Rigorous rationalism' (*strenger Rationalismus*), on the other hand, was especially useful precisely at the point where it established the boundaries and defined the limits of rational knowledge.[34]

On the psychological level 'pseudorationalism', he pointed out, had its roots in the same inscrutable aspirations which led men in former time to superstition, belief in a transcendent god, and faith in the prognostications of oracles, in order to fulfil their need to invest with certainty the decisions they had in fact taken themselves. Subsequently, in times of increasing cultural enlightenment, 'pseudorationalism' had come to form a surrogate (*Surrogat*) for the traditional sources of certainty, being the attempt to equip arguments devoid of secure foundations with the trappings of rationality (*Rationalität zu suggerieren*) by means of recourse to the tools of rhetorical suggestion (*Suggestionsapparate*).[35]

In two slightly later works, 'Prinzipielles zur Geschichte der Optik'[36] and 'Zur Klassifikation von Hypothesensystemen',[37] we find Neurath adopting a by now wholly conscious conventionalistic approach together with a historical and sociological perspective on the analysis of the structure and development of scientific thought.[38] Referring fully once again to Duhem and Poincaré and also to Whewell, he had no doubts about the hypothetical nature of scientific knowledge. Theories were hypotheses, and science was composed at any given time of a plurality of interconnected theoretical hypotheses. In choosing between competing or conflicting hypotheses, rules of 'correctness' (*Richtigkeit*) or of the greater or smaller explanatory power of theories did not normally obtain.[39] Rather, criteria of a philosophical, sociological, or pragmatic nature were more often at work.

For example it was quite possible, he argued, for the success of one theoretical system rather than another to depend on the spread (*Tragweite*) of its hypotheses and on the extent of the agreement of such hypotheses with others connected to them and themselves associated in their turn with general visions of the world prevalent at any given moment of time.[40] In rejecting a certain hypothesis or alternatively in retaining it by recourse to 'daring auxiliary hypotheses' (*gewagte Hilfshypothesen*)[41] to explain a particular fact, it was important for scientists, in making their decision, to consider whether other hypotheses were implicated in that decision. Thus, the wider the spread of a new theory or the wider the area of its interplay with other theories of wide spread, the more difficult it would be for this theory to succeed, as it would quickly find itself in conflict with general philosophies which met with universal accord.

Also, he said, it happened that certain promising theories were quickly abandoned – as with the theory of fluid electricity or with flogiston – simply because of the natural eagerness among scientists of a succeeding generation to change the theories they had inherited from their predecessors.[42] In other cases theories did not yield results and were abandoned because, as a result of the accidents of history, they found insufficient support, so that the potential concealed in them was not recognised until a much later stage and failed to be exploited at the right time.[43] Finally it was possible, as indeed was frequently the case, that competing theories were so different that comparison between them was difficult or that different theories were supported by different mathematicians and scientists with an equally persuasive degree of skill and adroitness.[44] In these, as in other cases, selection became more a matter of opportunity: criteria of practical utility were employed which fostered the establishment of those methodological 'routines' of a partly practical and a partly theoretical nature, to which the scientists of any given school attuned themselves,[45] or even criteria of purely practical convenience,[46] such as greater simplicity or ease of explanation were used:

in the much more transparent field of physics Poincaré, Duhem and others have amply shown that more than one self-consistent system of hypotheses can satisfy a given set of acts. If one physicist decides to adopt one system of hypotheses, and another a second, then a quarrel between them *cannot be settled* on the basis of the known facts. Certain inclinations and dispositions may have influenced the choices. If contradictions arise between physicists, then, if the facts are the same, we must ask whether these contradictions are due to *logical mistakes* or due to differences which *cannot be eliminated* by logic. Each of the two can say: "My hypothesis is self-consistent, compatible with the facts, and more pleasing to me".[47]

Finally, he argued – with reference once again to Duhem and Poincaré but deliberately exceeding the bounds of their limited conventionalism – it was not sufficient in the analysis of theories of physics to restrict oneself to consideration solely of their mathematical structures and of the explicitly discursive lines of their argumentation; it was necessary to take account also of the 'mental-image' component (*Phantasiebilder*) of theoretical hypotheses, i.e. of the extent to which philosophical theories depend on the 'images' (*Bilder und Bildchen*) employed in their construction. In certain historical cases this 'mental-image' component had played a part greater even than that of the logico-mathematical element.[48] And this, he observed, was a philosophical problem left open by the conventionalism of Duhem and Poincaré who had confined themselves to the identification of mathematical differences between theories of physics.[49]

It was, however, in his *Anti-Spengler*, a brilliant and precise critique of the methodological contradictions of Spengler's cosmography, that Neurath presented the first complete formulation of his 'scientific holism'.[50] Here too we find the first version of his famous metaphor of the development of scientific knowledge as the reconstruction of a ship in mid-ocean, of which he was later to make further use in his criticism of Carnap's understanding of his *Protokollsätze*:

We are like sailors who have to reconstruct their ship on the open sea but are never able to start afresh from the bottom. Where a beam is taken away a new one must at once be put there, and for this the rest of the ship is used as support. In this way, by using the old beams and driftwood, the ship can be shaped entirely anew, but only by gradual reconstruction.[51]

Any methodological basis of knowledge, he argued in this work, however much it aimed to free itself of its initial assumptions and interpretations, simply could not start, as Descartes had claimed, from an initial *tabula rasa*.[52] Its very terms, concepts, definitions, and linguistic interconnections conditioned from the outset, in a circular fashion which did not admit of being ignored, every subsequent conceptual development. Each subsequent development changed the centre of gravity of the whole structure of conceptual relations of which the totality of the preceding insights was composed, and each individual concept itself was influenced to some degree by this process of transformation. In the wake of each theoretical development previous conceptions of the world acquired an 'altered meaning' (*geänderter Sinn*) and, strictly speaking, demanded complete reformulation and re-exposition on the basis of these new meanings, despite the impossibility of ever arriving at a definitive position or of achieving absolute conceptual clarity.[53]

An idea, he explained, is like the flash of a torch on the dark walls of a tunnel: it shows us details of one restricted area at a time and, by collecting the data we receive from successive flashes in different areas, we are able to form an impression of the whole, even though we cannot see everything at once or make out with any clarity the surroundings of each area as it is illuminated. Each new idea is linked in with a web of concepts (*mit einem ganzen Begriffsnetz*) – the connection of flashes on the wall – and cannot be separated from it: the web emerges as a whole out of the dark as a plan providing some rough connection between individual illuminated areas, but not as a general illumination in itself. It was necessary therefore, as Duhem had shown, to recognise that all propositions relating to empirical phenomena were "impregnated with hypotheses of every sort" (*durchtränkt mit Hypothesen aller Art*) and that the web of hypotheses was derived from general conceptions of the world (*Weltanschauungen*) which were antecedent to scientific inquiry.[54]

NOTES

[1] Cf. H. Neider's testimony in 'Memories of Otto Neurath', *ES*, pp. 48–9. See also R. Haller and H. Rutte, 'Gespräch mit Heinrich Neider, Wien: Persönliche Erinnerungen an den Wiener Kreis', pp. 21 ff.
[2] Cf. H. Neider, *ES*, pp. 47–9, 82–3.
[3] R. Carnap, 'Intellectual Autobiography', pp. 24–9.
[4] Cf. K.R. Popper's testimony in 'Memories of Otto Neurath', *ES*, pp. 51–6.
[5] O. Neurath, *Wissenschaftliche Weltauffassung: Der Wiener Kreis*, Wien: Artur Wolf, 1929, now in *GpmS*, p. 304, English trans. *ES*, p. 305. "The university teachers I had met so far – H. Neider recounts in his testimony – had been epistemologists and logicians. History, sociology, the history of science and mathematics were alien to them. Neurath, on the other hand, was a man of encyclopedic knowledge – hardly any field was alien to him and he had published in many. His bookshelves were filled with works by scientists, philosophers, poets, fathers of the church. There were theological reference books and publications on current affairs" (*ES*, p. 46).
[6] O. Neurath, *Wissenschaftliche Weltauffassung: Der Wiener Kreis*, [1929], *GpmS*, p. 304, *ES*, p. 305. Neurath's manifesto ends with the following proposition, in italics: *"Die wissenschaftliche Weltauffassung dient dem Leben und das Leben nimmt sie auf"* (Ibid., *GpmS*, p. 315, *ES*, p. 318).
[7] Ibid., *GpmS*, pp. 301–5, *ES*, pp. 301–5.
[8] Ibid., *GpmS*, pp. 314–5, *ES*, p. 317.
[9] Ibid., *GpmS*, pp. 307–8, *ES*, pp. 308–9.
[10] Ibid., *GpmS*, p. 315, *ES*, p. 317.
[11] Ibid., loc. cit.
[12] See chap. 6, paragraph 3, below.
[13] In his 'Introduction' to *Modern Science and its Philosophy* Philipp Frank reminds us that Abel Rey's book *La théorie physique chez les physiciens contemporains*, Paris: Félix Alcan, 1907, soon translated into German (*Die Theorie der Physik bei den modernen Physikern*, by R. Eisler, Leipzig: W. Klinkhardt, 1908), played an important role in his discussions at Vienna with Otto Neurath and Hans Hahn in 1908–12. On the influence exerted by classical conventionalism over the 'first' Vienna Circle, see R. Haller, 'Il primo Circolo di Vienna', pp. 45–6. For the general repudiation of conventionalism by the members of the Vienna Circle, particularly by Schlick, Reichenbach and Carnap, see H. Feigl's authoritative testimony in 'The Origin and Spirit of Logical Positivism', pp. 3–24, now also in H. Feigl, *Inquiries and Provocations*, pp. 21–37. According to Feigl "Schlick, Reichenbach, and Carnap, though highly impressed with Poincaré's genius, repudiated his conventionalism. [...] Quite generally, what seemed correct to the logical positivists in the views of Poincaré and Duhem, is no more than the obvious logical truth that, given a theory in the form of a conjunction of postulates, and an empirically refuted consequence of that conjunction, any member of the conjunction may be false" (p. 32).
[14] O. Neurath, 'Zur Theorie der Sozialwissenschaften', in *Jahrbuch für Gesetzgebung, Verwaltung und Volkswirtschaft im Deutschen Reich, 34* (1910), pp. 37–67, now in *GpmS*, pp. 23–46.
[15] Neurath quotes the following books: P. Duhem, *La théorie physique. Son objet, sa*

structure, Paris: Marcel Rivière, 1906 (German trans. *Ziel und Struktur der physikalischen Theorien*, ed. by F. Adler, Leipzig: Johann Ambrosius Barth, 1908, with a preface by E. Mach; English trans. *The Aim and Structure of Physical Theory*, Princeton: Princeton University Press, 1954); H. Poincaré, *La science et l'hypothèse*, Paris: Ernest Flammarion, 1902 (German trans. *Wissenschaft und Hypothese*, Leipzig: Druck und Verlag von B.G. Teubner, 1906; English trans. *Science and Hypothesis*, London and Newcastle-on-Tyne: Walter Scott, 1905). In the *Literaturhinweise*, published as an appendix to the Vienna Circle's manifesto, Neurath quotes in addition: H. Poincaré, *La valeur de la science*, Paris: Flammarion, 1904 (German trans. *Der Wert der Wissenschaft*, Leipzig: Teubner, 1906); P. Duhem, *L'evolution de la mécanique*, Paris, Joanin, 1903 (German trans. by Philipp Frank: *Die Wandlungen der Mechanik und der mechanischen Naturerklärung*, Leipzig: Johann Ambrosius Barth, 1912).

[16] O. Neurath, 'Zur Theorie der Sozialwissenschaften', *Jahrbuch für Gesetzgebung, Verwaltung und Volkswirtschaft im Deutschen Reich, 34* (1910), now also in *GpmS*, p. 45.

[17] Ibid., loc. cit.

[18] Ibid., p. 46. Neurath refers to: W.S. Jevons, *The Principles of Science*, London: Macmillan, 1877; K. Pearson, *The Grammar of Science*, London: Adam and Charles Black, 1900; F. Enriques, *I problemi della scienza*, Bologna: Zanichelli, 1906 (English trans. *Problems of Science*, Chicago and London: Open Court, 1914).

[19] O. Neurath, 'Zur Theorie der Sozialwissenschaften', [1910], *GpmS*, p. 44.

[20] Ibid., loc. cit.

[21] Ibid., pp. 44–5.

[22] Ibid., p. 46.

[23] Ibid., p. 45.

[24] Ibid., p. 46.

[25] O. Neurath, 'Die Verirrten des Cartesius und das Auxiliarmotiv', *Jahrbuch der Philosophischen Gesellschaft an der Universität zu Wien 1913*, pp. 45–59, now in *GpmS*, pp. 57–67, English trans. *PP*, pp. 1–12. Cf. R. Haller, 'Das Neurath-Prinzip. Grundlagen und Folgerungen', pp. 81–2.

[26] O. Neurath, 'Die Verirrten des Cartesius und das Auxiliarmotiv', [1913], *GpmS*, p. 57, *PP*., p. 1.

[27] Ibid., *GpmS*, pp. 58–9, *PP*, pp. 2–3.

[28] Ibid., *GpmS*, p. 59 ("Auch das Denken bedarf der provisorischen Regeln in mehr als einer Hinsicht"), *PP*, p. 3.

[29] Ibid., *GpmS*, p. 60, *PP*, p. 4.

[30] Ibid., *GpmS*, p. 59 ("Wer eine Weltanschauung oder ein wissenschaftliches System schaffen will, *muß mit zweifelhaften Prämissen operieren*"), *PP*, p. 3.

[31] Ibid., loc. cit.

[32] Ibid., *GpmS*, p. 60, *PP*, p. 3.

[33] Ibid., loc. cit.

[34] Ibid., *GpmS*, p. 64, *PP*, p. 8.

[35] Ibid., loc. cit.

[36] O. Neurath, 'Prinzipielles zur Geschichte der Optik', *Archiv für Geschichte der Naturwissenschaften und der Technik, 5*, (1915), pp. 371–89, now in *GpmS*, pp. 71–84, English trans. *ES*, pp. 101–12.

[37] O. Neurath, 'Zur Klassifikation von Hypothesensystemen', *Jahrbuch der Philosophischen Gesellschaft an der Universität zu Wien 1914 und 1915*, pp. 39–63, now in *GpmS*, pp. 85–101, English trans. *PP*, pp. 13–31.

[38] It should be pointed out that according to Neurath research concerning the history of science had to allow not only for the logical structure of theories but also for the 'psychology of the researcher' (*die Psychologie der Forscher*): cf. 'Prinzipielles zur Geschichte der Optik', [1915], *GpmS*, p. 71.

[39] Ibid., loc. cit.

[40] Ibid., pp. 71–2.

[41] Ibid., p. 71.

[42] Ibid., p. 82.

[43] Ibid., pp. 82–3. R. Hegselmann stresses the importance and originality of these theses of Neurath's: cf. 'Otto Neurath. Empiristischer Aufklärer und Sozialreformer', pp. 40–1.

[44] O. Neurath, 'Prinzipielles zur Geschichte der Optik', [1915], *GpmS*, p.74.

[45] Ibid., p. 72.

[46] Ibid., p. 84.

[47] Otto Neurath, *Anti-Spengler*, München: Georg D.W. Callwey, 1921, now in *GpmS*, p. 188, English trans. *ES*, p. 203.

[48] O. Neurath, 'Prinzipielles zur Geschichte der Optik', [1915], *GpmS*, p. 73.

[49] Ibid., loc. cit.

[50] O. Neurath, *Anti-Spengler*, [1921], *GpmS*, particularly pp. 182–96, *ES*, pp. 197–213. On Spengler's philosophy of history see, most recently, W. Dray, *Perspectives on History*, London: Routledge and Kegan Paul, 1980, pp. 99–124.

[51] O. Neurath, *Anti-Spengler*, [1921], *ES*, p. 199 ("Wie Schiffer sind wir, die auf offenem Meer ihr Schiff umbauen müssen, ohne je von unten auf frisch anfangen zu können. Wo ein Balken weggenommen wird, muß gleich ein neuer an die Stelle kommen, und dabei wird das übrige Schiff als Stütze verwendet. So kann das Schiff mit Hilfe der alten Balken und angetriebener Holzstücke vollständig neu gestaltet werden, aber nur durch allmählichen Umbau", *GpmS*, p. 184).

[52] Ibid., *GpmS*, p. 183, *ES*, p. 198.

[53] Ibid., *GpmS*, pp. 183–4, *ES*, p. 198.

[54] Ibid., *GpmS*, p. 184, *ES*, p. 199.

CHAPTER 3

LINGUISTIC REFLEXIVITY AND 'PSEUDORATIONALISM'

1. METHODOLOGICAL DECISION AND THE REFLEXIVITY OF SCIENTIFIC LANGUAGE

The Vienna Circle's 1929 'manifesto' was dedicated to Schlick and involved, as we have noted, the participation of Carnap. Little more than a year later Neurath opened his philosophical attack on each of these two. Earlier than this, however, in 1928, he had published in the journal of the Austrian Social Democratic Party, 'Der Kampf', a joint review of two works by Carnap: *Der logische Aufbau der Welt* and *Scheinprobleme in der Philosophie*.[1] There he portrayed Carnap as an exponent of Viennese 'empirical rationalism' fighting a minority battle against theology and metaphysics (and especially Kantianism) in the university schoolrooms, a battle initiated by the adherents of French enlightenment and materialism and more recently taken further by such scientists as Mach, Avenarius, Poincaré, Russell, Reichenbach and Schlick.[2]

The philosophical movement represented by the *Wiener Schule* deserved, he claimed, to meet with sympathy from Austrian Marxists. Its critique of theology and metaphysics in the name of conceptual clarity and linguistic precision constituted an advance and a stimulus in the direction of Marxist theory and was, above all, a viable and effective contribution towards a 'general enlightenment' (*allgemeine Aufklärung*) in university and popular education. He voiced his doubts, however, about Carnap's attempt to elaborate a 'closed world-vision' (*geschlossenes Weltbild*) on logico-mathematical bases alone and to construct to this end an 'ideal language' (*ideale Sprache*).[3] Carnap had not taken account, he argued, of the insuperable difficulties involved in such a philosophical undertaking, probably for the reason that he concerned himself solely with the natural sciences and especially with physics. Had he extended his inquiry so as to include also the social sciences, he would have been more circumspect (*vorsichtiger*) and the question would have been raised of how development of scientific knowledge could be possible in cases where it was necessary to make use, in a 'varied intermixture' (*in bunter Mischung*), of both pure and impure conceptual instruments. 'Pureness' (*Sauberkeit*) of logical constructs had the advantage of guaranteeing the univocality (*Eindeutigkeit*) of formal theories. But an explanation was still needed as to how it was possible – indeed *whether* it was possible – to overcome the plurivocality (*Vieldeutigkeit*) obtaining in other areas of research – in the social sciences, for example, where the inherent necessity to make choices prevented 'univocality' and opened up a wide field to ideologies.[4]

In a slightly later work, 'Wege der wissenschaftlichen Weltauffassung', which marked the beginning of his collaboration on *Erkenntnis*,[5] Neurath brilliantly outlined a historical approach to the problems of science apparently in violent contradiction to the epistemological position of the Viennese group. Here not only did his language omit all positivist or scientific emphasis, not only did he propose the complete restoration of ancient magic into the tradition of western scientific thought, but also, surprisingly, he took as a premise of *wissenschaftliche Weltauffassung* so sober and pragmatic a conception of the possibilities and the bounds of scientific knowledge as to come close to outright scepticism.[6]

He placed very firm emphasis on the historical, sociological, ideological, and linguistic conditions which make the ways of science complex rather than linear, and also on the partial and contingent nature of all scientific knowledge – partiality and contingency confirmed by the breakdown of Newtonian causalism and the establishment, through Mach and Einstein, of a finitistic and relativistic view of science. Changes in modes of thought (*Denkweisen*), he argued, were closely linked to the actual technical and social transformations of mankind and did not follow any linear or unitary course or have any cumulative propensity, as a 'naively empiricist conception of history' would have us believe.[7]

Ancient magic, for example, was in some ways closer to the modern empirical sciences than the theological and metaphysical doctrines by which it had itself been replaced.[8] Theology and metaphysics laid claim to a wholesale explanation of the world by their reference to transcendent entities or to principles which were, by definition, subject to no possible empirical test, and they remained entirely untouched and unaltered by the developments of technology. Magic, on the other hand, operated in a manner which could in fact be defined as empirical: either in the sense that it laid no claim to explain the totality of the world, its origin or its destiny, or because it was actively engaged in dealings, rituals, or predictions which, like modern physics or biology, had distinct practical aims.[9] Even if the premises behind the explanations and predictions offered by ancient magic appeared today, from the modern standpoint, to be wrong, the same could not always be said for its conclusions. We owed to astrology, for example, the notion, bitterly contested at the time but now upheld as correct, that the flow of tides was connected with the motion of the moon.[10]

In many ways the modern scientist, whether engineer or doctor or sociologist, was nothing other than a 'magician' or 'astrologer' operating on different theoretical premises and making systematic use of logic and mathematics. This did not, however, allow modern science to be contrasted with ancient magic as the truth with error: our thought, he wrote,

> is an instrument of work (*Werkzeug*) and is dependent on our social relations and historical conditions. We ought never to forget that we cannot play at one and the same time the parts of defendant, prosecuting lawyer, and judge. We compare our actual experiences with the culture of the past, but we have no chance of making definitive judgments from a neutral position. Experimental proof of sentences on the basis of empirical procedures is part itself of a specific method.[11]

The ineradicable reflexivity of empirical knowledge, in so far as it was conditioned by its own methods of research and testing, made science dependent on contexts beyond its own narrow scientific sphere, and these contexts also had influence on the choice of methods, construction of theories, and their historical effect. This could be a matter of general conceptions of the world, such as Newton's, which involved the scientist's personal religious faith;[12] or it could be a matter of the particular historical course of a nation's culture, as with the development of science, from Lobatschewskij to Pavlov, in a 'young nation' like Russia;[13] or it could mean the very symbolic or lexical structure of a language: Chinese ideograms formed a serious obstacle to the free use of symbolic forms and consequently to the emergence of abstract disciplines, while a language like Arabic, for all its richness, had difficulty in accommodating metaphysical problems such as *Sein* and *Sosein*, which the German language – however infelicitously – fostered as a result of the wide freedom it allowed in the formation of words.[14]

But scientific knowledge, he continued, in addition to being conditioned in its development by extrascientific factors, was subject also to limitative elements of a logical and methodological type. Following the breakdown of Newtonian causalism and Laplacean determinism, scientific explanation now proceeded from one detail to another, "deducing from certain specific areas arguments for other specific areas",[15] without further reference to absolute parameters and without postulating 'laws of nature' of a strictly universal type and of a causal nature. Einstein's theory of relativity took account of and developed Mach's critique of the Newtonian conception of absolute time and space:[16] in opposition to this 'theological residue', Mach "has always held that physics describes, case by case, only partial aspects of the empirical processes and does not ever pass judgment on the 'world as a whole'".[17] In Mach's view, for example,

> given as it was to establishing functional relationships between observable phenomena, it was entirely natural that the principle of inertia should be ascribed to the system of fixed stars, rather than to absolute and infinite space.[18]

Further confirmation, he maintained, of the collapse of the causalist system founded on 'universal laws of nature' was to be seen in the introduction of statistical analysis into the field of physics itself. Having produced its first results, thanks to Quételet's work, in the area of sociology, statistical generalisation was now having the effect in modern physics of undermining the "absolute and comprehensive causal explanation, often strongly imbued with metaphysics".[19] Instead of the prediction of individual events conceived as the necessary effects of universal laws, scientific practice was now tending to accept as sufficient for its purposes probable hypotheses relating to classes or sequences of events.[20]

The indeterminacy and plurivocality of this type of analysis, whether it derived from an insufficient knowledge of the phenomena or from their intrinsic nature,[21] found its resolution in the determinacy and univocality of a practical *decision*, as when, for instance, one decided to assume the validity of a certain statistical law. But the univocality of this decision lay outside the ambit of strict scientific reasoning: no logical justification could be given for highly discretional choices such as the acceptance of a certain parameter of statistical probability (*Möglichkeitsmenge*).[22]

It was essential, he warned, for followers of *wissenschaftliche Weltauffassung* to become fully conscious of these difficulties and limitations to knowledge (*Enge und Begrenzheit der Erkenntnis*). Otherwise there was a risk that the postulation of absolute certainty and precision might lead to the creation of a new idol (*ein neues Idol*), destined simply to assume the place of the old *a priori* beliefs in a divinity and the infinite. In place of the priest or philosopher we would then have the professor.[23]

Wissenschaftliche Weltauffassung was not put forward as a new *Weltanschauung*. Neurath saw as essential the clear contradistinction between a philosophical view of the world as a whole ('*-anschauung*') and a scientific conception ('*-auffassung*') which used the term *Welt* not as a closed totality

(Totalität, abgeschlossenes Ganzes), but as the "area of scientific research in a state of daily evolution and expansion".[24] The scientific conception of the world therefore started from a twofold realisation: first, that it was not legitimate to consider human knowledge by reference to a totality of the world seen as independent of it, and, second, that this knowledge was, at the same time, subject to restrictions and limitations. *Wissenschaftliche Weltauffassung* therefore reaffirmed the dictum of the sophist Protagoras, at once both vaunting and humble, that "Man is the measure of all things".[25]

2. THE 'PHYSICALIST' OVERTURNING OF THE CIRCLE'S ORTHODOXY

Such were the premises, worked out over a period of twenty years, upon which, in 1930 and 1931, Neurath moved to his attack on the official *Erkenntnislehre* of the Vienna Circle with a critique aimed no less at the logical and linguistic constructivism of Carnap than at the epistemological realism of Schlick and of Wittgenstein's *Tractatus*.

The epistemological doctrine formulated by Neurath and published with the characteristic zeal that ensured it vast international repercussions, was 'physicalism'. Quite apart from its constructive element, this was a doctrine which turned out to be even more effectual from the point of view of the criticism it contained. Physicalism introduced to the heart of the Circle's epistemological debate a conception of science which was incompatible with the premises of logical empiricism, including Popper's falsificationism. It was Neurath's physicalism which sparked off what Passmore has called the 'insuperable internal difficulties' which were chiefly responsible for the Circle's demise.[26] Paradoxically it was Neurath's critique which – even more than Carnap's weak self-criticisms and certainly more effectively than the polemic of Popper (who, as is well known, claimed for himself the credit of having 'killed' logical positivism)[27] – engendered the rapid exhaustion of the very philosophical movement of which, on an organisational level, he continued himself to be the most active member.

Clearly Neurath's physicalism, in its first formulation, shared certain general philosophical premises with the Circle's orthodox *Erkenntnislehre* and, at least in its initial phase, did not take active issue with these. Among them was the conviction that only the empirical sciences were in a position to give relevant answers to 'meaningful' questions, so denying the possibility of attributing any competence or 'meaning' to philosophical thought which placed itself either above or to the side of scientific enquiry.[28] Also held in common was the idea of a clean theoretical division between science and 'metaphysics', a distinction to be seen in the tenacious attempt to find a criterion of formal and, if possible, linguistic and terminological 'demarcation' between on the one hand scientific sentences instilled with 'meaning' and on the other 'metaphysical' sentences devoid of any logical or empirical significance and hence of 'meaning'.[29] Lastly they shared the idea of the primacy of physics over all other means of cognition.

In accordance with these shared views Neurath saw the spatio-temporal co-ordinates as providing a procedural guarantee of the scientific and non-metaphysical nature of theories and propositions.[30] This position was taken to an extreme, however, by his physicalism, to the point where it ended as the paradoxical idea that the language itself of 'unified science' should be seen as a 'physical structure' (*physikalisches Gebilde*),[31] being conceived of as a succession of linguistic events positioned in time and space. But the idea turned out to be as sterile as it was paradoxical and justly received no significant further development in Neurath's thought. In other areas, however, and especially in three which were central to Viennese *Erkenntnislehre*, the contrast between Neurath's position and the main premises of logical empiricism could not have been more marked.

3. LANGUAGE AND REALITY: A METAPHYSICAL RELATIONSHIP

A major aspect of physicalism was the support it lent – by holding that it is not possible for language to be transcended – to a coherent and complete nominalism. Any attempt to transcend the sphere of language by means of language raised once more the traditional epistemological and metaphysical conundrums about the problem of the correspondence (*Übereinstimmung*) between language and the 'real world' (*Wirklichkeit, die wahre Welt*),[32] leading to the opposite and equally unacceptable results offered by idealism and epistemological realism.[33]

When we use speech, Neurath acutely argued in opposition to Wittgenstein,

> it is not possible for us to assume positions which are to some extent outside speech and so to be at one and the same time defendant, prosecutor, and judge.[34]

Our ability to speak of speech was, he pointed out, due only to the circular use of one part of language to describe and discuss other parts of language. The attempt, therefore, to "formulate propositions about the totality of language from a standpoint which is to a certain extent not yet linguistic", as Wittgenstein and some members of the Circle had claimed, was both profitless and contradictory.[35] While using linguistic means we could not, as it were, move into and then out of language in search of an extralinguistic foundation beneath linguistic statements.[36] Nor was it possible for us, in attempting to establish the 'truth' of linguistic hypotheses, to 'verify' them, as Schlick and the Circle had argued, by reference to 'things' or to 'reality'.[37] Nor did it make sense to talk of 'verification through immediate data' (*Verifikation durch ein Gegebenes*).[38]

From the theoretical point of view, science, no less than language, operated within the restraint of the reflexivity of propositions: for these constituted 'its point of departure and its point of arrival'.[39] Scientific propositions could be refuted only by other propositions. When a new scientific proposition was formulated – he argued, laying the teaching of classical conventionalism under contribution once again –

> it is confronted with the totality of propositions already accepted. If the new proposition agrees (*stimmt überein*) with those already in existence, it is aggregated with them; if it does not agree, it is defined as 'untrue' and is dismissed. Or, alternatively, a decision is made to modify the ensemble of established propositions in a way which permits the introduction of the new proposition.[40]

But, he added, such a decision was usually resorted to only with reluctance (*schwer*), since it required the alteration of the entire matrix of propositions accepted up to that point.[41] All the same, no other concept or criterion existed, beyond this reflexive concept of the truth, for establishing the 'correctness' of theoretical assertions. And this was sufficient to invalidate the notions of 'truth', or 'verification', or 'significance' (*Bedeutung*) espoused by members of the Circle.[42]

Here, then, Neurath rejected outright one of the cardinal principles of the neopositivist 'commonsense background' – the verificationist theory of meaning. In place of it he was himself to develop a pragmatic and historical conception of language and its meanings, close to that proposed by Wittgenstein in his late phase, and which he was to advance in opposition not just to the theses of Schlick and Carnap, but also to those of Popper and of Russell.[43]

We see here also the foundations laid by Neurath of his opposition to Tarski's formalisation of the semantic concept of 'truth', which was to be received, on the other hand, with enthusiasm by Carnap,[44] Hempel,[45] and even by Popper, who had no hesitation in hailing it, in homage to his own ontologism, as a 'theory of absolute and objective truth'.[46] Neurath was to be the first to identify the philosophical ambiguity of Tarski's formalisation of the 'intuitive' concept of truth and to force the author to attempt an important, albeit in all probability not wholly successful, clarification of it.[47] Tarski's formalisation (the sentence 'the snow is white' is true if and only if the snow is white) seemed to him ambiguous first on account of its explicit reliance on the Aristotelian conception of truth as 'correspondence to reality',[48] second on account of its logicistic assumption of empirical hypotheses within the binary scheme of classical logic,[49] and third on account of the apparent failure of Tarski's semantics to note that every empirical formulation was 'relative to a certain language' and that, because of the pluralism, variability, and imprecision of all languages, including artificial ones,

> a meeting (*Gegenüberstellung*) will never be possible between linguistic statements and reality, between thought and being, between cognition and the real world, between subject and object, between logical form and experience.[50]

More generally, as we shall see later on, 'semantics' in the logicistic form developed by Tarski and Carnap was to be for Neurath at most a means for the formal assessment of statements, but in no sense applicable, without specific auxiliary assumptions, to 'empirical arguments'. To fail, he said, to take account of this was to fall into the 'ontological fallacy' which rendered absolute, beyond their pragmatic and communicative function, the propositions and argumentative procedures of science.[51]

4. REFLEXIVITY AND THE GROWTH OF SCIENCE

What function was there left after this for the 'observation sentences' *(Beobachtungaussagen)* which in logical positivism formed the very basis of the empirical nature of science? Had 'observation' of facts any remaining significance, once direct comparison between scientific language and 'real' phenomena had been removed?

For Carnap, as already for Schlick in his *Positivismus und Realismus* and for Wittgenstein in his *Tractatus*, the whole structure of science was built on certain elements which guaranteed it cognitive 'security'. In the *Tractatus* science received its guarantee of certainty from the correspondence between 'atomic facts' and 'atomic propositions': every basic sentence was a representation (*Abbildung*) of a corresponding basic state of affairs (*Sachverhalt*). The sentence was true if such a state existed, and false if not, with the result that description of all the 'true' basic sentences formed a complete description of the world.[52] For Schlick, even before his formulation of the doctrine of the *Konstatierungen*,[53] 'certainty' sprang in a not dissimilar fashion from the possibility of establishing in the last analysis the meaning of linguistic symbols on the 'ostension' of the objects which they denoted and therefore of establishing the meaning of scientific propositions through their 'verification on the basis of immediate data'.[54]

For Carnap – both in the phenomenalistic and 'solipsistic' position which he took in his *Aufbau* and in successive works, such as 'Die physikalische Sprache als Universalsprache der Wissenschaft' and 'Psychologie in physikalischer Sprache',[55] which showed the influence of Neurath's physicalism – the 'certainty' of scientific knowledge lay in the concept of reducing the conceptual 'structure' to an 'immediate datum'. In his *Aufbau* (which drew on the phenomenalistic and reductionistic system of Russell's *Our Knowledge of the External World*)[56] the absolute quality of the 'immediate datum' appeared in the idea of the original material of cognition being formed from individuals' real experiences (*Erlebnisse*), in their simple and unadorned form, on the basis of which the *Konstitutionssystem* of 'fundamental relations' was to be built.[57]

In the two successive works, reference to the extralinguistic datum of real experience was discarded, and the absolute quality of the 'immediate datum' was transformed into the 'originality' of the 'protocol statements' elaborated 'beneath the guiding hand of experience' (*an Hand der Erfahrung*). In this first formulation by Carnap of the doctrine of the protocols, the idea of the recording of the observable characteristics of physical objects appeared in 'direct' and 'original' (*ursprünglich*) form, and so laid the foundations (*Grundlage*) for the objectivity and intersubjective communicability of the protocol language (*Protokollsprache*) of science.[58] For Carnap the protocol statements were immediately true and were not in themselves in need of any further confirmation (*die selbst nicht einer Bewährung bedürfen*).[59]

Neurath's physicalism was decisively opposed to each of these three fundamentalistic and reductionistic variants of empiricism, for which some years later Popper was to coin the term 'psychologism' (including even Neurath in this).[60] In 'psychologism' perceptive experience, as immediate cognition, provided the 'empirical base' which made the structure of science firm. It was in opposition to this 'psychologistic' system that Neurath issued the strong affirmation that

> there is no way to formulate pure protocol statements definitively taken as true and as

the basis of departure for science. No *tabula rasa* is possible.[61]

Here it was that Neurath presented a concise and pithy version of his famous nautical metaphor which was taken up and employed by many others, from Einstein[62] to Quine,[63] to express the idea of a *non-dogmatic* empirical attitude:

> we are like sailors who are forced to alter the structure of their ship in mid-ocean, with no possibility of dismantling her in harbour or of constructing her anew with better materials.[64]

In Neurath's view therefore, science was not to be envisaged as a structure erected on more or less solid foundations, although still resting on 'reality'. For him there was a circular movement in the growth of science; the metaphorical image which was better suited to expressing this reflexivity was that of a moving and floating structure, which, in order to continue afloat, was in constant need of reconstruction, but without being able to put into any dock or to use any materials other than those which could be salvaged from the old structure which needed the repair. In this circular process of gradual and continual reconstruction everything was *historically* modifiable but, at the same time, nothing could be totally modified:

> the process of the transformation of science consists in the fact that propositions employed during one period of time are discarded in the course of a successive period and other propositions are often substituted for them.[65]

If it sometimes happened that certain linguistic symbols survived over a great distance of time, that was due to their definition being in some way altered.[66]

The fate of elimination could befall also those observation records or 'protocols' which Carnap saw as immediate and 'original'. In fact even the 'protocols' submitted to the rule – in Rudolf Haller's term, '*das Neurath-Prinzip*' – by which 'concordance' (*Einklang, Übereinstimmung*) was seen as the general reason for the capacity of new scientific propositions to be assimilated into the corpus of already accepted propositions.[67] Nor could Carnap's thesis be accepted that, in the event of their incompatibility with the 'protocols',[68] it was only the 'laws' which could admit of modification. On the contrary, it might well be that in certain cases it would be more expedient to retain the laws untouched and to reject the observation sentences which were incompatible with them.

It was not only the case, therefore, that no proposition carried with it a definite *noli me tangere*, but it was also necessary to recognise that a scientific proposition was "definable precisely by virtue of the fact that it requires confirmation and that it is therefore itself eliminable".[69] Carnap, on the contrary, following in the tradition of Cartesian 'pseudorationalism', had attempted to establish absolutely certain and irrevocable bases for scientific knowledge.[70] In accordance with his 'methodological solipsism' he had seen the 'phenomenal' or 'experiential language' (*Erlebnissprache*) of private individuals as an 'original language'. He had failed to see that "every language is, as such, intersubjective" and that no "individual can assume only his own protocols as a basis of knowledge",[71] for the reason that the language through which people recorded their individual perceptive experiences preceded, and did not follow, the experiences themselves. Language therefore conditioned those perceptions and implicated them back to the intersubjective dimension intrinsic to linguistic communication, with the result that no 'originality', certainty, or purity was attainable in linguistic accounts of individual experience.[72]

In his revised version of the doctrine of 'protocols', despite surmounting his former 'methodological solipsism', Carnap continued to refer to observation sentences "which require no confirmation".[73] In this way, as Neurath critically observed, he failed to disassociate himself from the belief of traditional philosophical schools in 'immediate experiences' (*unmittelbare Erlebnisse*), schools which, with absolute confidence, laid down certain 'ultimate elements' (*letzte Elemente*), on the strength of which a fixed and unalterable conception of the world was attainable. 'Basic experiences' (*Atomerlebnisse*) were obviously not, in the light of such philosophical perceptions, susceptible to critical analysis and had no need of confirmation.[74] In exactly the same spirit, Neurath maintained, Carnap attempted to introduce a sort of 'basic protocol', in order to accommodate the need to distinguish clearly between 'direct' (*direkt*) observations, recorded by the protocol, and the subsequent 'elaborations' (*Verarbeitungen*) which the protocol statements underwent during the procedures of research.[75] He was therefore led to assume that any kind of 'elaboration' could be excluded from the protocol record. But that was an arbitrary assumption, because any protocol registration, although it used the intersubjective instrument of language, was always referable to an act of subjective perception and, as such, had to be referred to a specific individual forming the protocol.[76] It could not therefore have any originality against all other statements, and always carried within it some element of 'elaboration' on the part of the individual who recorded it.[77]

In the realm of scientific knowledge, therefore, no premise of certainty was provided by the protocols. Individual protocol statements did not, at base, differ from general sentences of a nomological kind arrived at through induction. Individual 'observations', no less than general principles reached inductively, were still the product of 'elaborations' and 'decisions'.[78]

For Neurath, therefore, no difference of principle existed between observation language and theoretical language. In this he not only preceded Popper's theory of the *Basissätze*, emptying it of originality, but also, surely, provided the inspiration for Quine's critique of the 'reductionist' dogma of empiricism.[79] The widespread consciousness of the *theory-ladeness* of all empirical observation, so characteristic of European and American philosophy of science in the Sixties and Seventies, thus turns out to have its direct precursor in Neurath.

5. THE PLURIVOCALITY AND IMPRECISION OF SCIENTIFIC LANGUAGE

One fundamental objective of logical empiricism, already present in Wittgenstein's *Tractatus* but pursued systematically by Carnap in the wake of Frege and Russell, was the logical analysis of the linguistic constructs employed in science and philosophy.[80] Closely linked with this was the aim also of building an exact scientific language, homogeneous and univocal, which by its very nature would find itself placed in categorical opposition to the language of metaphysics.

The new empiricism differed from classical empiricism – and especially from Mach – through its attempt to join two distinct and hitherto opposite strains of philosophical thought – logicism and empiricism.[81] It started from the basic premise that modern formal logic was to be seen as the domain of syntactically true or 'analytical' sentences.[82] Tautologies and contradictions, as the *Tractatus* instructed, were respectively true and false by virtue of basic logical principles which were self-evident and self-establishing, and were so, in Leibnizian terms, in respect of any possible world, logical constructs being semantically neutral. Formal logic was to be understood therefore not just as the conceptual instrument capable of providing a unitary and consistent basis for mathematics, but capable also of fulfilling the function, side by side with mathematics, of a general syntax of empirical science. The job of logic, once it had established a strict criterion of demarcation between science and metaphysics,[83] was to provide exact rules for the formalisation and transformation of the propositions and theorems of science. Thanks to the refined and powerful tools provided by modern formal logic, construction of a homogeneous language of science was possible in the form of an exact univocal language, free of all metaphysical contamination.[84]

As we have seen, Neurath fully shared the two neopositivistic aims of the confutation of metaphysics and the unification of science – indeed, within the Circle itself, he was the most committed and forceful proponent of these aims. It is all the more significant, therefore, that his contributions on both these themes were unfailingly critical towards Carnap's theses. The truth was that he nurtured the deepest distrust not only of Carnap's linguistic constructivism but also, more generally, of any evaluation which gave the analytical instruments of formal logic the position of only or overriding syntactical rules in empirical science. Furthermore, as we shall see, he tended to play down, as hardly relevant if not downright unfounded, the theoretical dichotomy between analytical and synthetic propositions, a dichotomy which was the backbone or, as Quine was to call it, fundamental dogma of Viennese orthodoxy.[85]

He did not deny altogether the usefulness of a linguistic therapy which would purge both ordinary and scientific language of 'metaphysical' terms and images, and he believed also that remedial criteria might, within certain limits, be provided by the spatio-temporal categories of physics. On the basis of a sort of selective grid he thought it possible that something approaching a lexical taxonomy could be set up, an *index verborum prohibitorum*, to work against metaphysics.[86] What he did oppose, however, was the idea that logical analysis of the syntactical structures of the language of science could ever lead to secure and definitive results, valid for all the sciences. He denied that logical analysis could provide a sufficient and in some ways automatic criterion for dispensing with metaphysics, because of a kind of logical incompatibility between science and metaphysics.

Advances in modern symbolic logic, he granted, had made possible a broad systematisation and axiomatisation of scientific theories. But he warned empiricists against 'complacency', since logic offered no 'certain guarantee against metaphysics'[87] and there was no possibility of constructing an ideal language on the basis of 'pure atomic propositions' (*saubere Atomsätze*). Such a hypothesis was as metaphysical as Laplace's fiction of the ghost.[88] Not even extensive and systematic use of logical symbolism could be considered as an approach to that ideal, because even expressions used in physics to refer to quantitative measures – e.g. "this thermometer registers 24 degrees" – contained elements of approximation and imprecision which could not be broken down.

Furthermore there existed no completely formalised system for the arrangement of scientific language which would allow it to be wholly separated from day-to-day language as it had historically evolved, containing within it an indefinite number of imprecise and unanalysed (*unanalysiert*) terms.[89] And then, even if it were possible to construct such a wholly scientific language, purged of all metaphysical components, it would be applicable "only to certain very specific sciences and perhaps only to parts of them".[90] In all other areas, he pointed out (to some extent in anticipation of late Wittgenstein),[91] the terms of ordinary speech and the terms of scientific language would of necessity continue to be intertwined in a mosaic of untraceable complexity. If one attempted to unravel the very last 'clusters' (*Ballungen*) of linguistic indeterminacy, one would discover that "imprecision may be reduced in one area but can always reappear with even greater force in another".[92]

But, as has been mentioned, Neurath appeared to raise also the deeper question of the nature (and importance) of the distinction between analytical and synthetic propositions. According to the 'orthodox' understanding, originated by Wittgenstein and taken up by Carnap, the distinction was one of principle: the properties of analytical propositions, belonging to logic and mathematics, were certainty and formal univocality, while synthetic propositions, which were empirically verifiable, were characterised by their reliance on factual situations. Neurath, while accepting Wittgenstein and Carnap's distinction between tautologies and empirical propositions, denied all the same that this distinction held on the level of the 'certainty' bases of sentences. What 'fundamentally' (*grundsätzlich*) distinguished an analytical, logical or mathematical, proposition from an empirical proposition was, he said, its 'univocality' (*Eindeutigkeit*), in the sense that a conflict between analytical propositions could be resolved through the application of previously agreed syntactical rules, which could not be the case with synthetic propositions. Here the multiplicity of (incompatible) propositions could be reduced only on the basis of a further methodological decision (*Entschluß*). But this did not, he argued, imply in the slightest that logical and mathematical propositions could be considered 'more secure' than those, for example, of optics:

> When we discuss chemical, biological or sociological theories, we ordinarily treat their logico-mathematical auxiliary tools as given and do not question them. This is only one form of a working technique and does not mean that we consider these mathematical tools as definitely secured. We cannot even pretend in a general way that we regard logico-mathematical proofs as more certain than statements of chemistry, biology or sociology. When we say that one statement is more certain than

another, we maintain something concerning our 'conduct' in this respect; for example that we do not intend to spend more time and effort in order to test its truth; moreover, that we do not foresee that the development of science must soon change it; in other words, what it would be necessary to do in this case, we do not feel obliged to do.[93]

Both analytical and synthetic propositions could, he said, in fact undergo modification at later stages on the basis of different circumstances or considerations. Contrary to the opinion of even the conventionalist Poincaré, who also opposed Russell's logicism, no absolute foundation existed for logic and mathematics, preserving them once and for all from the danger of contradiction. No mathematician could be absolutely certain that the formal constructs he employed did not involve, in some of their parts or in the course of their development, violations of logic.[94]

Here then we see Neurath very clearly invoking, against the logical and mathematical fundamentalism of Russell, Carnap, and Schlick the authority not only of Karl Menger[95] but also of an intuitionistic mathematician such as Luitzen Brouwer, who will have exercised a deep influence also on Wittgenstein in his *Philosophische Bemerkungen*, as well as on Gödel.[96] In *Mathematik, Wissenschaft und Sprache* Brouwer argued that even mathematics was conditioned by the irreducible imprecision of every language. No linguistic code existed which was so univocal as to exclude the possibility of misunderstanding and to guarantee, by virtue of its syntactical rigour, freedom from error. And this, said Brouwer, was true also for the attempt made by the formalist school to subject the language itself of mathematics to mathematical analysis. Far from advancing resolution of the problem, Hilbert's formalism had reversed it, so that it became a problem of metamathematical language.[97]

6. METHODOLOGICAL DECISION IN THE *PRAXIS* OF SCIENTIFIC COMMUNITIES

During these years Wittgenstein was living in Cambridge and was therefore distanced from Neurath's polemic. In all probability it had not even reached his attention. Carnap, on the other hand, characteristically took prompt note of the criticisms but without appreciating their full significance.[98] In his 'Über Protokollsätze' in a plainly self-critical vein, he credited Neurath with having been the first to expose the dangers of 'absolutism' present in logical positivism and, in order to allay the criticism, he now agreed that "absolute propositions of departure do not exist for the building of science"[99] and that it was in fact up to the individual researcher to decide which propositions he would actually take as 'protocol' points of departure.[100] He went further, however, when as a result of private conversations with Popper (whose unpublished manuscript of *Grundprobleme* he had recently read),[101] he expressed his belief that he could reconcile Neurath's views with those of Popper as being substantially the same. He ended by declaring himself in agreement with both, although he added a slight preference for Popper, who, to his mind, had gone 'a step further than Neurath'.[102] The greater simplicity of the theory of the *Basissätze*, understood as observation sentences themselves of a conventional and hypothetical nature on a par with all other theoretical propositions, may well account for this.[103] But here Carnap's principle of 'tolerance' only succeeded in confusing the issue by leaving important questions and distinctions untouched, as Popper himself was to claim later on.[104]

As it was, Carnap had proclaimed his agreement with the reasoning behind Neurath's conventionalism, but had made it into a simple question of 'free choice of the linguistic structures', while at the same time excluding the idea that 'convention' could concern 'the content of the synthetic propositions'. As he was to put it much later, in reply to the criticism of Robert S. Cohen, his "principle of tolerance in the logical syntax" did not contain "a doctrine of conventionally-chosen basic-truths". While recognising the conventional component in the language of science, he had intended to give "special emphasis to the non-conventional, objective component in the knowledge of facts" and stressed again that "the first operation in the testing of synthetic statements is the confrontation of the statement with observed facts".[105] It was a position, then, as weak on epistemological grounds as it was tacitly founded on an implicit epistemological realism, and one which was as far from the moderate conventionalism of Popper as it was from the more radical and consistent conventionalism advanced by Neurath.[106]

One critic who did achieve a clear understanding of the innovative character of the physicalistic programme and offered firm opposition to it was Moritz Schlick. In his 'Über das Fundament der Erkenntnis'[107] he made a clear distinction between the coherence theory of truth, which he attributed to Neurath, and the correspondence theory of truth, which gave science the task of "providing a *true* representation of the facts".[108] Only on the basis of a conception of the truth as the agreement between linguistic propositions and empirical reality – he maintained, introducing the doctrine of the *Konstatierungen* as synthetic non-hypothetical propositions – was it possible to arrive at 'absolutely certain' criteria in the construction and examination of scientific hypotheses.[109]

The contrary theory, he argued, by which internal coherence was made the sole condition of validity in scientific language, left it up to the scientist to decide whether or not to accept the results of empirical observation: it did not in fact provide any criterion for distinguishing an observation account of empirical facts from a series of internally logically compatible sentences of any kind, even the most far-fetched. Thus Neurath's empiricism came close, he claimed, to outright relativism and scepticism.[110]

Even Hempel, who entered the fray in support of Neurath, saw the contrast between Neurath and Schlick in terms of the choice between a doctrine of truth as coherence or as correspondence.[111] These, then, were the ways by which it came to be almost universally believed – by Ayer,[112] for instance, as well as Russell,[113] Popper,[114] Kraft,[115] Scheffler,[116] Passmore,[117] and Lakatos[118] – that physicalism was to be interpreted as a coherence theory of truth and as implicit epistemological realism, which Neurath forcefully repudiated on many occasions and for which no confirmation exists in any of his writings.[119]

So we find Ayer writing in his much-cited 'Verification and Experience':

> Neurath makes the truth and falsehood of any proposition whatsoever depend upon its compatibility or incompatibility with other propositions. He recognizes no other criterion. In this respect, his protocol propositions are not allowed any advantage. [...] All we have to do if we wish a proposition to be stable is to decide to accept it and to reject any proposition that is incompatible with it. The question whether such a decision is empirically justified or not is one to which, according to the implications of Neurath's doctrine, no meaning can be attached.[120]

A little further on we see the first appearance of the charge of implicit epistemological realism:

> [Neurath and Hempel] do not mean by a protocol proposition one which can be directly verified by observation, for they deny that this is possible. They use the term 'protocol' purely as syntactical designation for a certain assemblage of words. But why should one attach special significance to the *word* 'observation'? [...] There is no more justification for it than there would be for making a collection of all the propositions that could be correctly expressed in English by sentences beginning with the letter B, and choosing to call them Basic propositions. If Neurath and Hempel do not recognize this it is probably because, in writing about *Protokollsätze*, they unconsciously employ the forbidden criterion of agreement with experience.[121]

Bertrand Russell discussed Neurath and Hempel's theses on the protocol statements at some length in his 1940 Harvard lectures published under the title *An Inquiry into Meaning and Truth*, where he ended with some severe strictures on Neurath (to which Neurath replied, as we shall later see, with an open attack on Russell's 'ontologism'):

To say: "A is an empirical fact" is, according to Neurath and Hempel, to say: "the proposition 'A occurs' is consistent with a certain body of already accepted propositions". In a different culture circle another body of propositions may be accepted; owing to this fact, Neurath is an exile. He remarks himself that practical life soon reduces the ambiguity, and that we are influenced by the opinions of neighbours. In other words, empirical truth can be determined by the police. This doctrine, it is evident, is a complete abandonment of empiricism, of which the very essence is that only experiences can determine the truth or falsehood of non-tautologous propositions.[122]

And again:

Neurath's doctrine, if taken seriously, deprives empirical propositions of all meaning. When I say "the sun is shining", I do not mean that this is one of a number of sentences between which there is no contradiction; I mean something which is not verbal, and for the sake of which such words as 'sun' and 'shining' were invented. The purpose of words, though philosophers seem to forget this simple fact, is to deal with matters other than words. If I go into a restaurant and order my dinner, I do not want my words to fit into a system with other words, but to bring about the presence of the food. I could have managed without words, by taking what I want, but this would have been less convenient. The verbalist theories of some modern philosophers forget the homely practical purposes of every-day words, and lose themselves in a neo-neo-Platonic mysticism. I seem to hear them saying 'in the beginning was the Word', not 'in the beginning was what the word means.[123]

Following Popper, Imre Lakatos has recently written in the same vein:

Neurath rejects falsificationism as 'pseudo-rationalism'; but where is 'real rationalism'? Popper warned already in 1934 that Neurath's permissive methodology (or rather lack of methodology) would make science unempirical and therefore irrational: "We need a set of rules to limit the arbitrariness of 'deleting' (or else 'accepting') a protocol sentence. Neurath fails to give any such rules and thus unwittingly throws empiricism overboard [...] Every system becomes defensible if one is allowed (as everybody is, in Neurath's view) simply to 'delete' a protocol sentence if it is convenient". Popper agrees with Neurath that all propositions are fallible; but he forcefully makes the crucial point that we cannot make progress unless we have a firm rational strategy or method to guide us when they clash.[124]

What basis is there to these criticisms? There can be no doubt that Neurath's affirmation of the (linguistic) intranscendability of the language of science, as of all languages, demonstrated his opposition to a conception of 'truth' as a correspondence of scientific propositions to the 'facts' or to the 'real world' *(Wirklichkeit, wahre Welt)*. But he was at the same time equally distant from the coherence theory of truth which we have seen attributed to him by his commentators, to say nothing of the 'neo-neo-Platonic mysticism' gratuitously laid to his charge by Russell in the paragraph above. It is also hard to see how, with Ayer, Neurath was subject to an implicit or unconscious realism, because it is obvious that his methodological and 'terminological' physicalism is not to be confused with a philosophical materialism,[125] and also because Neurath's idea of language as a 'physical fact' should not be considered anything more than a fruitless and redundant attempt towards the circular establishment of the intrascendability of language – an attempt, admittedly out of place, to eradicate from physicalistic grammar the 'metaphysical' problem of the relation between propositions and 'reality'.[126]

What seems to have escaped the attention of the commentators – Scheffler in particular – is that Neurath's rigorous nominalism and conventionalism went hand in hand with an equally precise and lucid evaluation of scientific activity from the pragmatic and historical/sociological point of view. One might say that Neurath had no coherence theory of truth because he had no theory of truth. This was a conception which was to surface, decades later, only in such 'antipositivistic' historians and philosophers of science as Michael Polanyi, Hanson, Quine, and Kuhn, and finds perhaps its only real counterpart in Ludwik Fleck's doctrine of thought-style (*styl myślowy, Denkstil*) and of thought-collective (*kolektyw myślowy, Denkkollektiv*).[127]

Neurath made detailed reply in two works, 'Radikaler Physikalismus und 'Wirkliche Welt'' and 'Universal Jargon and Terminology', to the criticisms both of Schlick (and, by implication, of Ayer) and of Russell. He pointed out that to assimilate physicalism to a generic 'coherence theory of truth' was entirely to overlook its empirical aspect.[128] In physicalism, he emphasised, the point of departure in the elaboration of any theoretical hypothesis was to be found in the 'observation sentences' (*Beobachtungssätze*) of the protocols. Also the central goal of every theoretical hypothesis and the very condition of our being able to test it lay in forward predictions (*Voraussagen*), which could be seen, as he pointed out with consistent pragmatism, as 'the alpha and omega of science'. Predictions could in their turn only be tested on the basis of observation sentences.[129]

From this it is clear that Neurath, in just the same way as Schlick and the other exponents of logical positivism, saw empirical observation, recorded by the observation sentences, as an essential element of scientific activity. The deep and radical disagreement between them arose over the relationship between empirical observation and the 'observation sentences' on the one hand and, in the realm of induction, between the 'observation sentences' and theoretical hypotheses on the other. For Neurath, as we have seen, protocol recording was, although indispensable, a 'subjective elaboration' *(Verarbeitung)* with no specific connotation of 'immediacy', 'originality', or of absolute precision or certainty, as the methodological solipsism of Carnap or the linguistic subjectivism of Russell would claim.[130] The requirement of certainty, central to Schlick's theory of the *Konstatierungen*, confused with a problem of scientific logic (*Wissenschaftslogik*) a problem of research psychology, to be studied eventually in the context of a behaviouristic sociology of knowledge (*Behavioristik der Erkenntnis*).[131]

It was always possible, then, and indeed permissible, to leave a particular observation sentence out of account, especially when it conflicted with the totality of hypotheses already accepted up to that point, on the basis of which predictions had been made which had since been confirmed by other observation sentences.

And this held also for the inductive process leading from observation sentences to the formulation of hypotheses. Here too it was 'on the basis of decisions' (*auf Grund von Entschlüssen*),[132] and not as the result of logical procedures,[133] that certain observation sentences rather than others were taken as premises and as the parameters of testing for theoretical hypotheses. And, as had been demonstrated by Duhem and Poincaré as well as by the history itself of physics (and modern optics in particular), it was always as a result of methodological decisions that separate systems of hypotheses were constructed on the basis of the same observation sentences, though differing among themselves and equally capable of leading to empirically corroborated predictions.[134] It was necessary, therefore, to recognise that methodological decisions occupied – as they could hardly not – an extremely wide area in the work of scientists.[135] But to admit this, in particular to recognise that criteria of opportunity – such as the simplicity or coherence or predictive capacity of a hypothetical model – could prevail over (differing) accounts of empirical observation, was not tantamount, as Schlick, Ayer, and Russell had claimed, to opening the door to radical relativism. It was not the same as abandoning empiricism for a methodological version of idealism, in the form of a coherence theory of truth. In this reply, then, we see an example of Neurath at his clearest and firmest, with his mind at its most original in meeting head-on the opposition arising from orthodox neopositivism along with its non-inductivist variations.

As to the charge of relativism, Neurath rebutted this by pointing out how it ignored the historical and sociological dimension of the scientific process.[136] The presence of methodological decisions within the practical activity of science did not in any way introduce unaccountable elements of judgement, even if it did preclude the measurement of scientific procedures and results against 'an unalterable standard of truth'.[137] He did not see methodological decisions, as Victor Kraft[138] understood him too, as an act of arbitrary selection on the part of the individual researcher from an indefinite and unlimited number of possibilities. It was, on the contrary, a decision guided by practical rules and conditioned by a variety of factual circumstances which reduced the scope of the 'logically' available alternatives.

Above all, the point of decision could not be located so simply within the individual decision-maker: every scientist, he pointed out, worked within more or less wide groups of research, by whose approach the scientific thoughts of the individual members were strongly influenced and their freedom of choice circumscribed.[139] Research projects (*Pläne*) which were oriented towards specific scientific areas involved in themselves a reduction in the number of possible expectations and predictions.[140] Moreover, the individual researcher would master no more than one scientific system in the course of his life, and would hardly be in a position to set up adequate tests of the conditions of applicability of a multiplicity of scientific systems.[141] Even more to the point, the individual scientist working within each area of research found that he had to play down his personal theoretical outlook, however promising in itself, in order to operate according to those 'modes of thought' (*Denkweisen*) which met with greater agreement and support, and which would increase his personal chances of obtaining results which would be recognised as scientific.[142]

These factors provided an explanation, in Neurath's view, for the historical phenomenon of theoretical possibilities never being developed, in any very great number, by several different groups at once, and also of the contrary phenomenon of the general arrival, by processes of adaptation and selection *(Anpassung und Selektion)*, at a 'kind of co-ordination of whole generations of scientists' (*eine Art Angleichung ganzer Generationen*),[143] even without taking into account cases in which certain lines of scientific thought were proscribed or prohibited.[144]

If, therefore, the actual development of science was studied in a historical manner, we would find that the corpus of scientific discoveries owed its relative unity not to the logically necessary – and thus univocal – character of its theoretical advances, but rather to the 'praxis of life' (*Praxis des Lebens*).[145] Reduction of complexity in the scientific system was due, therefore, to historical factors which linked *practically* the decisions of scientists, as well the success of their theories,[146] to 'common conditions of social life and technological development'.[147] The process of the unification of scientific methodologies and practices was thus allied (*verwandt*) to the process of unification of production techniques, commerce, and war, with which it in many respects shared such strong connections.[148]

In the light of this clearly-defined 'externalist' position, the task of the historian or of the philosopher of science turned out not to be one of attempting to reconstruct the 'logic' of scientific knowledge on absolute and immutable bases: such bases did not exist and could not therefore be found. More simply and modestly, historians and philosophers needed to restrict themselves to ascertaining which scientific propositions were from case to case shared by certain research communities:

> We possess no fixed point which may be made the fulcrum for moving the earth; and in like manner we have no absolutely firm ground upon which to establish the sciences. Our actual situation is as if we were on board ship on an open sea and were required to change various parts of the ship during the voyage. We cannot find an absolute immutable basis for science; and our various discussions can only determine whether scientific statements are accepted by a more or less determinate number of scientists and other men. New ideas may be compared with those *historically* accepted by the sciences, but not with an unalterable standard of truth.[149]

The most rational and unitary reconstruction of scientific knowledge would, he said, therefore concern itself above all with the historical and sociological consideration of actual scientific praxis, and so of the opinions shared and the decisions taken by the scientific communities. We should not lose sight of the fact that "all the sciences, *qua* means of prediction, exist for the benefit of the development of life", and that the history of science "cannot be studied in isolation, but should be seen as the history of certain means within the overall context of the history of human existence".[150]

7. EMPIRICAL RATIONALISM AND 'PSEUDORATIONALISM'

When the historical and sociological conception of science was contrasted with the ontological and logicistic conception, two quite different models of rationality presented themselves to Neurath's mind: empirical rationalism on the one side and 'pseudorationalism' of the Cartesian type on the other.

The claim of 'pseudorationalism', Neurath maintained, was to build the edifice of science piece by piece on the 'sure ground' of precision, univocality of theories, and their correspondence to reality. The 'pseudorationalist' refused to acknowledge that the pieces of information available at any given historical moment were always limited and relatively imprecise and that there was consequently wide scope left for choice between different possibilities. Irrationally, therefore, he attempted to limit reason to logical univocality.[151] When a decision was to be taken on a specific question, his psychological condition was such that he could not be happy with careful consideration of a multiplicity of factors or reasons as the basis of his choice: he would prefer instead to be able to base the decision, if not on some justification of a transcendent kind, then at least on rigorous and univocal inferential logic.[152]

The 'pseudorationalist' treated every object whatever "as if it were entirely calculable" and took no account of the 'multiplicities of possible interpretations' of theories or of the part necessarily played in the actual practice of organised science by collectively shared decisions (*Gemeinsamkeit der Beschlüsse*).[153] He overlooked the fact that "at root the whole of science is in its totality always under discussion" and that even the terms which appeared in the simplest observation sentences, such as "Charles sees that the thermometer is registering 30 degrees", involved "the entire complex of science".[154]

Such a 'pseudorationalistic' attitude, in Neurath's view, did a disservice (*diskreditiert*) to empiricism.[155] Empiricism itself did not entail any metaphysical conception of science as knowledge of reality (*Wirklichkeitserkenntnis*), but this was exactly the claim of Russell's 'pseudorationalistic' ontologism. In levelling the charge of neo-Platonic mysticism against physicalism, Russell's critique, Neurath argued, was couched to a significant degree in terms of the 'existence' of reality, affirming the ontological pre-eminence of the objects denoted by language ("in the beginning was what the word means") and correspondingly denying the ontological pre-eminence of language. But this, Neurath claimed, was no more than a sterile problem of metaphysics which scientific discourse should be at pains to avoid.[156] He clarified his own position thus: in arguing for the non-transcendability of language, he had not intended to assert the pre-eminence of thought or of language in respect to 'being' or the 'objects' ("in the beginning was the Word"). What he had done was to formulate a methodological principle whose significance seemed totally to have escaped the 'ingenuous' realism practised by Russell: it was not 'facts' and 'propositions' which confronted one another in scientific research, but only propositions differing between themselves in their description of experiences or perhaps in their differing interpretations of the same experiences. It was misguided therefore to refer to a 'comparison' (*Vergleich*) with the facts as the solution to a difference between propositions.[157]

Furthermore, the relative nature of every scientific proposition could not be got over, as Russell claimed, by means of an appeal to a kind of empirical 'absolute truth in itself'. Even if such an absolute 'empirical' truth were to exist, it would still have to be expressed through the vehicle of language and share thereby in the limitations and historical/sociological variability of any language. It could not be opposed dogmatically to the 'error' of people who would not recognise it, but would have to be considered as just one opinion amongst others in the debates between 'different groups of thinkers'. And because the opinions of social groups were naturally subject to the direct and indirect influences of the social and political context in which they found themselves, they were indeed conditioned even by the 'police', to whom Russell referred with such misplaced irony in his attack.[158]

Thus, since it could not be a 'pseudorationalistic' doctrine of empirical 'truth', empiricism was, as Neurath saw it, sufficiently established in the actual work of scientists by the 'decision' to test hypotheses and predictions systematically on the basis of observation sentences. The 'rationality' of empiricism lay, then, in the aim of seeing that hypotheses and predictions coincided *to the greatest degree possible* with the accounts of observation.[159]

NOTES

1 Cf. *Der Kampf, 21* (1928), pp. 624–6, now in *GpmS*, pp. 295–7.

2 Ibid., *GpmS*, pp. 295–6.

3 Ibid., p. 296. On the relationship between Carnap's *Konstitutionssystem* and Neurath's physicalism see E. Runggaldier, 'Das Problem der 'Darstellung' bei Neurath und Carnap', in *Wittgenstein, der Wiener Kreis und der Kritische Rationalismus*, pp. 245–7.

4 Cf. O. Neurath, Review of R. Carnap, *Der logische Aufbau der Welt*, [1928a], *GpmS*, p. 296.

5 *Erkenntnis, 1* (1930–31), pp. 106–25, English trans. *PP*, pp. 32–47.

6 Cf. R. Haller, 'Geschichte und wissenschaftliches System bei Otto Neurath', in *Wittgenstein, der Wiener Kreis und der Kritische Rationalismus*, p. 306.

7 O. Neurath, 'Wege der wissenschaftlichen Weltauffassung', *Erkenntnis, 1* (1930–31), pp. 107–8, 116, English trans. *PP*, p. 33.

8 Ibid., p. 108, *PP*, pp. 33–4; cf.: O. Neurath, *Empirische Soziologie*, Wien: Julius

Springer, 1931, now in *GpmS*, pp. 426–35, English trans. *ES*, pp. 319–421; O. Neurath, 'Magie und Technik', *Erkenntnis, 2* (1931), pp. 529–31.
[9] O. Neurath, 'Wege der wissenschaftlichen Weltauffassung', [1930/31], pp. 108–10, *PP*, pp. 33–5.
[10] Ibid., p. 111, *PP*, p. 36.
[11] Ibid., p. 123, *PP*, p. 46.
[12] Ibid., p. 118, *PP*, p. 42.
[13] Ibid., p. 117, *PP*, p. 41.
[14] Ibid., p. 123, *PP*, p. 46.
[15] Ibid., p. 108, *PP*, p. 34.
[16] Ibid, loc. cit.; cf. O. Neurath, 'Physicalism: The Philosophy of the Viennese Circle', *The Monist, 41* (1931), p. 621, now also in *PP*, pp. 49–50.
[17] O. Neurath, 'Wege der wissenschaftlichen Weltauffassung', [1930/31], p. 119, *PP*, p. 42.
[18] Ibid., loc. cit.
[19] Ibid., p. 122, *PP*, p. 45; cf. O. Neurath, *Wissenschaftliche Weltauffassung: Der Wiener Kreis*, [1929], *GpmS*, p. 310, *ES*, p. 312.
[20] O. Neurath, 'Wege der wissenschaftlichen Weltauffassung', [1930/31], p. 122, *PP*, p. 45. On causationism, and the need for overcoming it, see the lively debate in *Erkenntnis, 1* (1930–31), with essays of H. Reichenbach, R. von Mises, P. Hertz, F. Waismann, H. Feigl. See also A. Pannekoek, 'Das Wesen der Naturgesetzes', ibid., *3* (1932–33), pp. 389–400.
[21] O. Neurath, 'Wege der wissenschaftlichen Weltauffassung', [1930/31], p. 122, *PP*, p. 45.
[22] Ibid., p. 123, *PP*, p. 45.
[23] Ibid., loc. cit.
[24] Ibid., p. 107, *PP*, p. 33.
[25] Ibid., p. 125, *PP*, p. 47.
[26] Cf.: J. Passmore, 'Logical Positivism', in *Encyclopedia of Philosophy*, vol. 5, p. 56; J. Passmore, *A Hundred Years of Philosophy*, London: Duckworth, 1966, pp. 367–93; K.R. Popper, 'Autobiography', pp. 69–71. See in addition R. Haller, 'New Light on the Vienna Circle', pp. 25 ff.
[27] Cf. K.R. Popper, 'Autobiography', pp. 69–70.
[28] O. Neurath, 'Physikalismus', *Scientia, 50* (1931), now also in *GpmS*, pp. 417–8, English trans. *PP*, p. 52; cf. O. Neurath, 'Soziologie im Physikalismus', *Erkenntnis, 2* (1931), p. 394, English trans. *PP*, p. 59.
[29] O. Neurath, 'Wege der wissenschaftlichen Weltauffassung', [1930/31], p. 116, *PP*, p. 40; O. Neurath, *Lebensgestaltung und Klassenkampf*, Berlin: E. Laub, 1928, now in *GpmS*, pp. 287–93, English trans. *ES*, pp. 290–8; O. Neurath, 'Soziologie im Physikalismus', [1931f], p. 393 ("alle sinnvollen Aussagen sind in den *Wissenschaften* enthalten"), *PP*, p. 58.
[30] O. Neurath, 'Physikalismus', [1931d], *GpmS*, p. 418, *PP*, p. 53.
[31] Ibid., loc. cit.
[32] Ibid., loc. cit.; O. Neurath, 'Radikaler Physikalismus und 'Wirkliche Welt'', *Erkenntnis, 4* (1934), p. 354, English trans. *PP*, p. 107; O. Neurath, 'Pseudo', [1935a], pp. 354, 364, English trans. *PP*, pp. 122, 130.
[33] O. Neurath, 'Physikalismus', [1931d], *GpmS*, p. 418, *PP*, p. 53.

[34] O. Neurath, 'Soziologie im Physikalismus', [1931f], p. 397, *PP*, p. 61.
[35] Ibid., p. 396 ("man kann sich nicht über die Sprache als Ganzes sozusagen von einem 'noch-nicht-sprachlichen' Standpunkt aus äußern"), *PP*, pp. 60–1; O. Neurath, 'Physikalismus', [1931d], *GpmS*, p. 418, *PP*, p. 52.
[36] Ibid., *GpmS*, p. 419, *PP*, p. 54.
[37] Ibid., *GpmS*, p. 418 ("So werden immer *Aussagen mit Aussagen* verglichen, nicht etwa mit einer 'Wirklichkeit', mit 'Dingen' wie es bisher auch der Wiener Kreis tat"), *PP*, p. 53.
[38] O. Neurath, *Einheitswissenschaft und Psychologie*, Wien: Gerold & Co, 1933, now in *GpmS*, p.594.
[39] O. Neurath, 'Soziologie im Physikalismus', [1931f], p. 397 ("Aussagen [sind] Ausgangspunkt und Ende der Wissenschaft"), *PP*, p. 61.
[40] O. Neurath, 'Physikalismus', [1931d], *GpmS*, p. 419, *PP*, p. 53; O. Neurath, 'Soziologie im Physikalismus', [1931f], p. 403, *PP*, p. 66.
[41] Ibid., loc. cit.
[42] Ibid., loc. cit. ("Die hier aufgestellte Definition von 'richtig' und 'unrichtig' entfernt sich von der im 'Wiener Kreis' üblichen, die aus 'Bedeutung' und 'Verifikation' rekurriert").
[43] Cf.: O. Neurath, 'Pseudo', [1935a], pp. 362–3, *PP*, p. 129; O. Neurath, 'Universal Jargon and Terminology', in *Proceedings of the Aristotelian Society*, New Series, *41*, 1941, pp. 144–8, now in *PP*, pp. 226–9 ("Our proposals lead to history and sociology of the sciences and to a stressing of the social implications of language", ibid., p. 229); O. Neurath, 'Radikaler Physikalismus und 'Wirkliche Welt'', [1934], p. 356 ("Die Termini 'Satz', 'Sprache' usw. müssen historisch-soziologisch definiert werden"), *PP*, p. 108.
[44] Cf. R. Carnap, 'Intellectual Autobiography', pp. 60–2. At the First International Congress for the Unity of Science, organised by Neurath in Paris (1935), Carnap took sides with Tarski against Neurath, who together with Årne Naess had criticized Tarski's paper 'Grundlegung der wissenschaftlichen Semantik' (see *Actes du Congrés International de Philosophie Scientifique*, Paris, 1935, *Actualités Scientifiques et Industrielles*, no. 290, Paris: Hermann & Cie Editeurs, 1936, III, pp. 1–8; cf.: R. Carnap, 'Wahrheit und Bewährung', ibid., IV, pp. 18–23; R. Carnap, 'Intellectual Autobiography', p. 61.
[45] Cf.: C.G. Hempel, 'Studies in the Logic of Confirmation', *Mind*, *54* (1945), p. 115; C.G. Hempel, 'Le problème de la vérité', *Theoria*, *3* (1937), pp. 225–8.
[46] Cf. K.R. Popper, 'Autobiography', pp. 70, 78, 114; K.R. Popper, 'Truth, Rationality, and the Growth of Scientific Knowledge', in *Conjectures and Refutations*, London: Routledge & Kegan Paul, 1969, pp. 223–7. Popper's interpretation of Tarski's semantic theory of truth has been criticized by S. Haack, in *Philosophy of Logics*, Cambridge: Cambridge University Press, 1978; see also T.S. Kuhn, 'Reflections on My Critics', in I. Lakatos, A. Musgrave (eds.), *Criticism and the Growth of Knowledge*, p. 265.
[47] Cf. O. Neurath, 'Erster Internationaler Kongreß für Einheit der Wissenschaft in Paris 1935', *Erkenntnis*, *5* (1935), pp. 395–402; A. Tarski, 'The Semantic Conception of Truth', in H. Feigl, W. Sellars (eds.), *Readings in Philosophical Analysis*, p. 80, note 3 (with reference to Neurath's criticism). Many others after Neurath have criticized Tarski's formalization of the semantic concept of truth: for instance F.

Gonseth, H. Reichenbach, F. Kaufmann, M. Black, G.C. Field and, more recently, H. Putnam, in *Meaning and the Moral Sciences*, Boston: Routledge & Kegan Paul, 1978, pp. vii, 1–6, 9–17, 28–32.

[48] O. Neurath, 'Einzelwissenschaften, Einheitswissenschaft, Pseudorationalismus', in *Actes du Congrès International de Philosophie Scientifique, Sorbonne, Paris 1935*, I, *Actualités Scientifiques et Industrielles*, no. 388, Paris 1936, now in *GpmS*, pp. 706–7, English trans. *PP*, pp. 135–6; O. Neurath, 'Zu den Vortragen von Black, Kokoszynska, Williams', *Erkenntnis*, *7* (1937–38) ("[...] ehe noch die Frage entschieden ist, innerhalb welcher Kalküle der 'Wahrheitsbegriff', wie ihn Tarski und andere jetzt vertreten, wertfoll ist, und ob er vielleicht gewisse absolutistische Existenzbehauptungen in verschleierter Form nahelegt", ibid., p. 374), English trans. *PP*, p. 208; O. Neurath, 'Physikalismus und Erkenntnisforschung', I, *Theoria*, *2* (1936), now in *GpmS*, pp. 753–4, English trans. *PP*, p. 163; O. Neurath, *Foundations*, [1944a], pp. 12–3.

[49] Ibid., loc. cit. This aspect has been particularly emphasised by F. Kaufmann; see Carnap's reply to Kaufmann's criticism in *Introduction to Semantics*, Cambridge (Mass.): Harvard University Press, 1942; cf. H. Putnam, *Meaning and the Moral Sciences*, pp. 2–3.

[50] O. Neurath, 'Physikalismus und Erkenntnisforschung', [1936f], I, *GpmS*, p. 754, *PP*, p. 163.

[51] O. Neurath, *Foundations*, [1944a], pp. 12–3, 48, note 19.

[52] Cf. L. Wittgenstein, *Tractatus logico-philosophicus*, London: Routledge & Kegan Paul, 1961, 4.25, 4.26, 4.3.

[53] See C.G. Hempel, 'Schlick und Neurath: Fundierung versus Kohärenz in der wissenschfatlichen Erkenntnis'. See in addition: H. Rutte, 'Neurath kontra Schlick. Zur Wahrheitsdiskussion im Wiener Kreis', in *Wittgenstein, der Wiener Kreis und der Kritische Rationalismus*, pp. 248–51; T. Oberdan, 'Neurath's Complaint', ibid., pp. 241–3; H. Rutte, *Wahrheit und Basis empirischer Erkenntnis*, Amsterdam: Rodopi, 1982.

[54] Cf. M. Schlick, 'Positivismus und Realismus', *Erkenntnis*, *3* (1932–33), pp. 7, 29.

[55] In *Erkenntnis*, *2* (1931), pp. 432–65; ibid., *3* (1932–33), pp. 107–42.

[56] Cf. B. Russell, *Our Knowledge of the External World*, London: George Allen & Unwin, 1926. W.V.O. Quine has pointed out the relationship between the two books; cf.: W.V.O. Quine, 'Russell's Ontological Development', *Journal of Philosophy*, *63* (1966), pp. 107–42; W.V.O. Quine, *Ontological Relativity and Other Essays*, p. 74. For a recent reinterpretation of Carnap's *Aufbau*, with references to Neurath and Quine's ontological relativism, see A. Coffa, 'Idealism and the Aufbau', in N. Rescher (ed.), *The Heritage of Logical Positivism*, Lanham (Md.): University Press of America, 1985, pp. 133–55.

[57] R. Carnap, *Der logische Aufbau der Welt*, Berlin and Schlachtensee: Weltkreis Verlag, 1928, p. 105, English trans. Berkeley: University of California Press, 1967.

[58] R. Carnap, 'Die physikalische Sprache als Universalsprache der Wissenschaft', *Erkenntnis*, *2* (1931), pp. 437, 438.

[59] Ibid., p. 438.

[60] Cf. K.R. Popper, *Logik der Forschung*, pp. 51–55, English trans. pp. 93–7. See also G. Andersson, 'How to Accept Fallible Test Statements? Popper's Criticist Solution', in G. Andersson (ed.), *Rationality in Science and Politics*, Dordrecht and Boston: D.

Reidel, 1985, pp. 47–65.
[61] O. Neurath, 'Protokollsätze', *Erkenntnis, 3* (1932–33), p. 206 ("*Es gibt kein Mittel, um endgültig gesicherte saubere Protokollsätze zum Ausgangspunkt der Wissenschaften zu machen.* Es gibt keine tabula rasa"), English trans. *PP*, p. 92.
[62] Cf. R. Carnap, 'Intellectual Autobiography', p. 38.
[63] Cf.: W.V.O. Quine, 'Identity, Ostension, and Hypostasis', now in W.V.O. Quine, *From a Logical Point of View*, p. 79 ("The philosopher's task was well compared by Neurath to that of a mariner who must rebuild his ship on the open sea"); W.V.O. Quine, *The Ways of Paradox and Other Essays*, p. 210 ("As Neurath has said, we are in the position of a mariner who must rebuild his ship plank by plank while continuing to stay afloat on the open sea"), pp. 212–3 ("Epistemologists have wanted to posit a realm of sense data, [...] for fear of circularity [...]. But if with Neurath we accept this circularity, simply recognizing that the science of science is a science, then we dispose of the epistemological motive for assuming a realm of sense data"); W.V.O. Quine, *Word and Object*, p. 3 ("Neurath has likened science to a boat which, if we are to rebuild it, we must rebuild plank by plank while staying afloat in it. The philosopher and the scientist are in the same boat."), pp. 123–4 ("The interlocked conceptual scheme of physical objects, identity and divided reference is part of the ship which, in Neurath's figure, we cannot remodel save as we stay afloat on it. The ontology of abstract objects is part of the ship too, if only a less fundamental part. The ship may owe its structure partly to blundering predecessors who missed scuttling it only by fools' luck. But we are not in a position to jettison any part of it, except as we have substitute devices ready to hand that will serve the same essential purposes"), p. 210; W.V.O. Quine, *Ontological Relativity and Other Essays*, pp. 83–4 ("This interplay is reminiscent again of the old threat of circularity, but it is all right now that we have stopped dreaming of deducing science from sense data. We are after an understanding of science as an institution or process in the world, and we do not intend that understanding to be any better than the science which is its object. This attitude is indeed one that Neurath was already urging in Vienna Circle days, with his parable of the mariner who has to rebuild his boat while staying afloat in it"), pp. 126–7 ("I see philosophy and science as in the same boat – a boat which, to revert to Neurath's figure as I so often do, we can rebuild only at sea while staying afloat in it. There is no external vantage point, no first philosophy").
[64] O. Neurath, 'Protokollsätze', [1932/33], p. 206 ("*Wie Schiffer sind wir, die ihr Schiff auf offener See umbauen müssen, ohne es jemals in einem Dock zerlegen und aus besten Bestandteilen neu errichten zu können*"), *PP*, p. 92.
[65] O. Neurath, 'Protokollsätze', [1932/33], p. 208, *PP*, p. 94.
[66] Ibid., loc. cit. This thesis has been developed by I. Lakatos with reference to fundamental concepts of mathematics in *Proofs and Refutations. The Logic of Mathematical Discovery*, Cambridge: Cambridge University Press, 1976.
[67] O. Neurath, 'Protokollsätze', [1932/33], pp. 209–10, *PP*, pp. 95–6. Cf. R. Haller, 'Das Neurath-Prinzip. Grundlagen und Folgerungen', passim. C.G. Hempel, in 'Schlick und Neurath: Fundierung versus Kohärenz in der wissenschaftlichen Erkenntnis', pp. 8–9, emphasises the importance of this 'principle' and recognizes its historical primacy with respect to Carnap and Popper's similar theses.
[68] O. Neurath, 'Protokollsätze', [1932/33], p. 210, *PP*, p. 95.
[69] Ibid., loc. cit.

[70] According to J. Passmore "Reichenbach once wrote of Carnap that 'his theory may be regarded, after a fashion, as a modern fulfilment of Descartes' quest for an absolutely certain basis of science'. In so far as that was his ambition, most logical positivists would now admit, he certainly failed to fulfil it" (*A Hundred Years of Philosophy*, p. 393).
[71] O. Neurath, 'Protokollsätze', [1932/33], pp. 211–2, *PP*, pp. 96–7. "If someone makes predictions and wants to check them himself, he must count on changes in the system of his senses, he must use clocks and rulers; in short the man who supposedly is in isolation already makes use of the 'intersensual' and 'intersubjective' language. The forecaster of yesterday and the controller of today are, so to speak, two persons" (O. Neurath, 'Physikalismus', [1931d], *GpmS*, p. 420, *PP*, p. 55).
[72] O. Neurath, 'Protokollsätze', [1932/33], pp. 211–2, 213, *PP*, pp. 96–7, 98; cf. E. Stiffler, 'The Vienna Circle and the Pragmatic Conception of Scientific Protocols', in *Wittgenstein, der Wiener Kreis und der Kritische Rationalismus*, pp. 252–5.
[73] O. Neurath, 'Protokollsätze', [1932/33], pp. 206–8, 210–1, *PP*, pp. 93–4, 95–6.
[74] Ibid., p. 210, *PP*, p. 96.
[75] Ibid., loc. cit.
[76] Ibid., pp. 210–2, *PP*, pp. 96–7.
[77] Ibid., p. 210, *PP*, p. 96.
[78] O. Neurath, 'Physikalismus', [1931d], *GpmS*, p. 418, *PP*, p. 53.
[79] Cf. W.V.O. Quine, 'Two Dogmas of Empiricism', *Philosphical Review*, *60* (1951), pp. 20–43, now in W.V.O. Quine, *From a Logical Point of View*, pp. 20–46. And yet Quine seems not to recognise the importance of Neurath's contribution on this subject: cf. W.V.O. Quine, *Ontological Relativity and Other Essays*, p. 85.
[80] Cf.: R. Carnap, 'Die neue und die alte Logik', *Erkenntnis*, *1* (1930–31), pp. 12–26; R. Carnap, 'Überwindung der Metaphysik durch logische Analyse der Sprache', ibid., *2* (1931), pp. 219–41. On the relationship between Wittgenstein's philosophy and Neurath's thought see F. Wallner, 'Wittgenstein und Neurath. Vergleich von Intentionen und Denkstil', in *Schlick und Neurath. Ein Symposion*, ed. by R. Haller, pp. 419–23.
[81] O. Neurath, *Le développement du Cercle de Vienne et l'avenir de l'empirisme logique*, Actualités Scientifiques et Industrielles, no. 290, Paris: Hermann & Cie Editeurs, 1936, pp. 10–2; O. Neurath, *Wissenschaftliche Weltauffassung: Der Wiener Kreis*, [1929], *GpmS*, pp. 305–8, *ES*, pp. 305–10.
[82] R. Carnap, 'Die neue und die alte Logik', pp. 21–3; see the discussion between Carnap and Quine in the Fifties: W.V.O. Quine, 'Carnap and Logical Truth', in P. A. Schilpp (ed.), *The Philosophy of Rudolf Carnap*, pp. 385–406; R. Carnap, 'W.V.O. Quine on Logical Truth', ibid., pp. 915–22. See also W.V.O. Quine, 'On Carnap's Views on Ontology', *Philosophical Studies*, *2* (1951), n. 5, pp. 65–72.
[83] Cf. K.R. Popper, 'The Demarcation between Science and Metaphysics', in P.A. Schilpp (ed.), *The Philosophy of Rudolf Carnap*, pp. 183–226, and Carnap's reply 'K.R. Popper and the Demarcation between Science and Metaphysics', ibid., pp. 877–81.
[84] R. Carnap, 'Die neue und die alte Logik', pp. 15, 23–6; R. Carnap, 'Überwindung der Metaphysik durch logische Analyse der Sprache', pp. 221–33; R. Carnap, 'Die physikalische Sprache als Universalsprache der Wissenschaft', pp. 432–3.
[85] Cf. W.V.O. Quine, 'Two Dogmas of Empiricism'; see also M.G. White, 'The

Analytic and the Synthetic: An Untenable Dualism', in L. Linsky (ed.), *Semantics and the Philosophy of Language*, Urbana: The University of Illinois Press, 1952, pp. 272–86.

[86] O. Neurath, *Anti-Spengler*, [1921], *GpmS*, p. 192, *ES*, p. 208; O. Neurath, 'Protokollsätze', [1932/33], p. 205, *PP*, pp. 91–2; O. Neurath, *Einheitswissenschaft und Psychologie*, [1933], *GpmS*, p. 506.

[87] O. Neurath, *Le développement du Cercle de Vienne et l'avenir de l'empirisme logique*, [1936], pp. 56–7.

[88] O. Neurath, 'Protokollsätze', [1932/33], p. 204, *PP*, p. 91.

[89] Ibid., loc. cit.

[90] Ibid., p. 205, *PP*, p. 91.

[91] Cf. A. Gargani, *Introduzione a Wittgenstein*, Roma and Bari: Laterza, 1973, pp. 79–91.

[92] O. Neurath, 'Protokollsätze', [1932/33], p. 206 ("Wird die Unpräzision an einer Stelle verringert, kann sie wohl gar an anderer Stelle verstärkt wieder auftreten"), *PP*, pp. 92 f.

[93] O. Neurath, 'L'encyclopédie comme 'modèle'', *Revue de Synthèse*, *12* (1936), n. 2, p. 188, English trans. *PP*, p. 146; O. Neurath, 'Radikaler Physikalismus und 'Wirkliche Welt'', [1934], pp. 350–1, *PP*, pp. 103–4.

[94] Ibid., p. 351, *PP*, p. 104; O. Neurath, 'L'encyclopédie comme 'modèle'', [1936d], pp. 188–9, *PP*, pp. 146–7.

[95] O. Neurath, 'Radikaler Physikalismus und 'Wirkliche Welt'', [1934], p. 351, *PP*, p. 104. Neurath quotes Karl Menger's essay 'Die neue Logik', in K. Menger, *Krise und Neuaufbau in den exacten Wissenschaften*, Leipzig und Wien: Deutike, 1933, pp. 107, 119.

[96] Cf.: A. Gargani, *Introduzione a Wittgenstein*, pp. 44–7; R. Haller, 'Rede zur Eröffnung des Schlick-Neurath-Symposion 1982', in R. Haller (ed.), *Schlick und Neurath. Ein Symposion*, p. xx.

[97] Neurath quotes L.E.J. Brouwer's essay *Mathematik, Wissenschaft und Sprache*, Wien, Wiener Gastvorträge 1928, also in *Monatshefte für Mathematik und Physik*, *36* (1929), pp. 153–264. For a criticism of Carnap, Hempel and Reichenbach's philosophy of mathematics see S.F. Barker, 'Logical Positivism and the Philosophy of Mathematics', in P. Achinstein and S.F. Barker (eds.), *The Legacy of Logical Positivism*, pp. 229–57.

[98] Popper and Agassi critically emphasize this aspect of Carnap's intellect; see: K.R. Popper, 'The Demarcation between Science and Metaphysics', in P.A. Schilpp (ed.), *The Philosophy of Rudolf Carnap*, p. 184; J. Agassi, Review of J. Hintikka (ed.), *The Secret of Carnap: Rudolf Carnap, Logical Empiricist*, Dordrecht and Boston: D. Reidel, 1975, in *Philosophia*, *10* (1981), pp. 57–63.

[99] R. Carnap, 'Über Protokollsätze', *Erkenntnis*, *3* (1932–33), pp. 215–8.

[100] Ibid., p. 224.

[101] In 'Über Protokollsätze' Carnap refers to Popper's opinions and theses, which he knew thanks to some private conversations he had with Popper during common holidays in the Tyrol, and also through his knowledge of Popper's unpublished typescript *Die beiden Grundprobleme der Erkenntnistheorie*; cf. K.R. Popper, 'Autobiography', pp. 65–7.

[102] R. Carnap, 'Über Protokollsätze', p. 228.

[103] Ibid., pp. 223–4. Cf.: K.R. Popper, 'Replies to My Critics', in P.A. Schilpp (ed.), *The Philosophy of Karl Popper*, p. 969; K.R. Popper, 'Autobiography', p. 71.
[104] Ibid., pp. 969–70; K.R. Popper, 'The Demarcation between Science and Metaphysics', p. 211, note 60.
[105] R. Carnap, 'Robert S. Cohen on the Relationship between Dialectical Materialism and Empiricism', in P.A. Schilpp (ed.), *The Philosophy of Rudolf Carnap*, p. 864.
[106] For Carnap's implicit epistemological realism see F. Barone, *Il neopositivismo logico*, pp. 217–23.
[107] Cf. *Erkenntnis, 4* (1934), pp. 79–99.
[108] Ibid., pp. 84 ff. On the relationship between Schlick and Neurath see: K. Lehrer, 'Schlick and Neurath: Meaning and Truth', in R. Haller (ed.), *Schlick und Neurath. Ein Symposion*, pp. 49–61; R. Hilpinen, 'Schlick on the Foundation of Knowledge', ibid., pp. 63–78; W. Leinfellner, 'Schlicks kognitive Erkenntnistheorie als Wissenschaftstheorie', ibid., pp. 103–27; H. Lauener, 'Neurath's Protocol Sentences and Schlick's 'Konstatierungen' versus Quine's Observation Sentences', ibid., pp. 129–48; H. Tennessen, 'Qualms about Otto Neurath's Cabby Language', ibid., pp. 385–98; J. Dvorak, 'Wahrscheinlichkeit, Logik und Empirie: Neurath, Schlick, und Zilsel', ibid., pp. 465–70; D. Davidson, 'Empirical Content', ibid., pp. 471–89; O. Hanfling, *Logical Positivism*, Oxford: Basil Blackwell, 1981, pp. 82–6.
[109] Cf. *Erkenntnis, 4* (1934), pp. 91–9.
[110] Ibid., pp. 86–7.
[111] Cf.: C.G. Hempel, 'On the Logical Positivists' Theory of Truth', *Analysis, 2* (1935), pp. 49–59; C.G. Hempel, 'Some Remarks on 'Facts' and Propositions', ibid., pp. 93–6; C.G. Hempel, 'Some Remarks on Empiricism', ibid., *3* (1936), pp. 30–40; see also: B. Juhos, 'Kritische Bemerkungen zur Wissenschaftstheorie des Physikalismus', *Erkenntnis, 4* (1934), pp. 397–418; T. Vogel, 'Bemerkungen zur Aussagentheorie des Radikalen Physikalismus', ibid., pp. 160–4. In his recent essay, 'Schlick und Neurath: Fundierung versus Kohärenz in der wissenschaftlichen Erkenntnis', p. 17, Hempel recognises that Neurath's thesis cannot be understood as 'eine reine Kohärenzauffassung'.
[112] Cf.: A.J. Ayer, 'Verification and Experience', in A.J. Ayer (ed.), *Logical Positivism*, Glencoe (Ill): The Free Press, 1959; A.J. Ayer, *The Foundations of Empirical Knowledge*, London: Macmillan, 1940.
[113] Cf. B. Russell, *An Inquiry into Meaning and Truth*, London: Allen & Unwin, 1950.
[114] K.R. Popper, *Logik der Forschung*, pp. 54–5, English trans. pp. 95–7.
[115] Cf. V. Kraft, 'Popper and the Vienna Circle', in P.A. Schilpp (ed.), *The Philosophy of Karl Popper*, pp. 185–201, particularly p. 194.
[116] Cf. I. Scheffler, *Science and Subjectivity*, Indianapolis: Bobbs-Merrill, 1967, pp. 91–123.
[117] Cf. J. Passmore, *A Hundred Years of Philosophy*, p. 376 ("Thus Neurath's attack on metaphysics leads him back to that coherence theory of truth already familiar to us in the writings of the Absolute Idealists – not surprisingly, since for them, too, 'transcendence' was the great enemy").
[118] Cf. I. Lakatos, 'Falsification and the Methodology of Scientific Research Programmes', pp. 113–4.
[119] In his reply to R.S. Cohen's criticism ('Robert S. Cohen on the Relationship

between Dialectical Materialism and Empiricism') Carnap maintains that Neurath's formulation of the theory of non-transcendability of linguistic propositions was 'misleading', but it did not imply any 'coherence theory' of truth. (It is, by the way, worth mentioning that, while defending Neurath from the charge of nominalism, Carnap seems to be inclined to attribute to him a sort of epistemological realism, precisely akin to his own implicit realism). According to Carnap Neurath's formulations "were indeed repeatedly interpreted in this sense, not only by outsiders like Russell and Ayer, but also by Schlick. Neurath vehemently rejected this interpretation in the discussions of the Vienna Circle, and also in a remark in his report on the Paris Congress of 1935 (*Erkenntnis*, *5*, 1936, p. 400). At any rate, there cannot be any doubt that Neurath never held this conception" (P.A. Schilpp, ed., *The Philosophy of Rudolf Carnap*, p. 864). As to Neurath's remark mentioned by Carnap, cf. O. Neurath, 'Erster Internationaler Kongress für Einheit der Wissenschaft in Paris 1935', [1935c], p. 400; see also O. Neurath, *Foundations*, [1944a], p. 47, note 16.

[120] A.J. Ayer, 'Verification and Experience', pp. 230–1.

[121] Ibid., p. 231.

[122] B. Russell, *An Inquiry into Meaning and Truth*, p. 148.

[123] Ibid., pp. 148–9.

[124] According to Lakatos Neurath's reply (in 'Pseudorationalismus der Falsifikation') to Popper's criticism shows that Neurath "never grasped Popper's simple argument" (cf. I. Lakatos, 'Falsification and the Methodology of Scientific Research Programmes', p. 114).

[125] In *Einheitswissenschaft und Psychologie*, *GpmS*, p. 591, Neurath openly maintains the identity between 'Physikalismus' and 'räumlichzeitliche Terminologie'.

[126] Cf. F. Barone, *Il neopositivismo logico*, pp. 310–1, 314; G. Statera, *Logica, linguaggio e sociologia. Studio su Otto Neurath e il neopositivismo*, pp. 48–9.

[127] Cf. L. Fleck, *Entstehung und Entwicklung einer wissenschaftlichen Tatsache. Einführung in die Lehre vom Denkstil und Denkkollektiv*, Basel: Benno Schwabe, 1935; see also: T.S. Kuhn, *The Structure of Scientific Revolutions*, Chicago: University of Chicago Press, 1970, p. x; T.S. Kuhn, 'Foreword' to L. Fleck, *Genesis and Development of a Scientific Fact*, Chicago and London: University of Chicago Press, 1979, pp. vii–xi; L. Schäfer and T. Schnelle, 'Einleitung' to L. Fleck, *Erfahrung und Tatsache*, Frankfurt a.M.: Suhrkamp Verlag, p. 347; R.S. Cohen and T. Schnelle (eds.), *Cognition and Fact. Materials on Ludwik Fleck*.

[128] O. Neurath, 'Radikaler Physikalismus und 'Wirkliche Welt'', [1934], p. 347, *PP*, p. 101.

[129] Ibid., p. 348, *PP*, pp. 101–2; O. Neurath, 'Physikalismus', [1931d], *GpmS*, pp. 418–9, *PP*, pp. 52–4; O. Neurath, 'Einheit der Wissenschaft als Aufgabe', *Erkenntnis*, *5* (1935), p. 17, English trans. *PP*, p. 116.

[130] Cf.: B. Russell, *An Inquiry into Meaning and Truth*, p. 139; O. Neurath, 'Universal Jargon and Terminology', [1941], pp. 146–8, *PP*, pp. 228–9.

[131] O. Neurath, 'Radikaler Physikalismus und 'Wirkliche Welt'', [1934], pp. 351, 359, *PP*, p. 104, 110.

[132] Ibid., p. 348, *PP*, p. 102.

[133] Ibid., p. 352 ("Unter den miteinender konkurrierenden Satzsystemen wählen wir ein aus. Dies so ausgewählte Satzsystem ist aber *nicht logisch ausgezeichnet*"), *PP*, p. 105; O. Neurath, 'Einheit der Wissenschaft als Aufgabe', [1935], p. 20 ("Aber solche

Eindeutigkeit des Entschlusses und der Tat ist *nicht* logisches Ergebnis aus irgendwelchen Prämissen, die zu *einer* einzigen Prognose über den Erfolg der Tat führen"), *PP*, p. 118.

134 O. Neurath, 'Radikaler Physikalismus und 'Wirkliche Welt'', [1934], p. 352, *PP*, p. 105.

135 O. Neurath, *Le développement du Cercle de Vienne et l'avenir de l'empirisme logique*, [1936], p. 57.

136 Cf.: O. Neurath, 'Radikaler Physikalismus und 'Wirkliche Welt'', [1934], p. 353 ("Aber 'unserer Standpunkt' ist dennoch nur historisch bestimmt"), *PP*, p. 106; ibid., p. 356 ("Die Termini 'Satz', 'Sprache' usw. müssen historisch-soziologisch definiert werden"), *PP*, p. 108; O. Neurath, 'Unified Science and Its Encyclopedia', *Philosophy of Science*, *4* (1973), p. 276, now also in *PP*, p. 181.

137 Ibid., loc. cit.

138 Cf. V. Kraft, 'Popper and the Vienna Circle', p. 194.

139 O. Neurath, 'Radikaler Physikalismus und 'Wirkliche Welt'', [1934], p. 352, *PP*, p. 105; O. Neurath, 'Einheit der Wissenschaft als Aufgabe', [1935], p. 19, *PP*, p. 117.

140 O. Neurath, 'Radikaler Physikalismus und 'Wirkliche Welt'', [1934], p. 352, *PP*, p. 105; O. Neurath, 'Einheit der Wissenschaft als Aufgabe', [1935], p. 20, *PP*, pp. 117–8.

141 O. Neurath, 'Radikaler Physikalismus und 'Wirkliche Welt'', [1934], p. 353, *PP*, p. 106; O. Neurath, 'Einheit der Wissenschaft als Aufgabe', [1935], pp. 18–9, *PP*, pp. 116–7.

142 Ibid., p. 19, *PP*, p. 117. The analogy between these elements of sociology of science and some central theses of Kuhn's *The Structure of Scientific Revolutions* may be found surprising; cf. R. Hegselmann, 'Otto Neurath. Empiristischer Aufklärer und Sozialreformer', pp. 40–1.

143 O. Neurath, 'Einheit der Wissenschaft als Aufgabe', [1935], p. 19, *PP*, p. 117.

144 Ibid., loc. cit.

145 O. Neurath, 'Radikaler Physikalismus und 'Wirkliche Welt'', [1934], p. 352, *PP*, p. 105.

146 O. Neurath, 'Einheit der Wissenschaft als Aufgabe', [1935], pp. 20, 21, 19, *PP*, pp. 117–8, 119, 117.

147 Ibid., p. 17, *PP*, p. 115.

148 Ibid., p. 21, *PP*, p. 119.

149 O. Neurath, 'Unified Science and Its Encyclopedia', [1937a], p. 276, *PP*, pp. 180–1.

150 O. Neurath, *Empirische Soziologie*, [1931e], *GpmS*, p. 426, *ES*, p. 319.

151 O. Neurath, *Anti-Spengler*, [1921], *GpmS*, p. 140, *ES*, p. 159.

152 O. Neurath, 'Pseudo', [1935a], p. 360, *PP*, p. 127.

153 O. Neurath, *Empirische Soziologie*, [1931e], *GpmS*, p. 517, *ES*, p. 407.

154 O. Neurath, 'Einheit der Wissenschaft als Aufgabe', [1935], p. 20, *PP*, p. 118.

155 Ibid., loc. cit.

156 O. Neurath, 'Universal Jargon and Terminology', [1941], p. 145, *PP*, p. 227.

157 Ibid., loc. cit.; O. Neurath, 'The Danger of a Careless Terminology', *The New Era*, *22* (1941), no. 7, p. 148.

158 O. Neurath, 'Universal Jargon and Terminology', [1941], p. 147, *PP*, pp. 228–9.

159 O. Neurath, 'Radikaler Physikalismus und Wirkliche Welt', [1934], pp. 356–7

("So reduziert sich für uns das Streben nach Wirklichkeitserkenntnis auf das Streben, die Sätze der Wissenschaft in Übereinstimmung zu bringen mit möglichst vielen Protokollaussagen. Das ist aber *sehr viel*: hierin ist der *Empirismus* begründet"), *PP*, p. 109.

CHAPTER 4

NEURATH VERSUS POPPER

1. POPPER'S CRITICISM OF NEURATH

In 'Pseudorationalismus der Falsifikation', an article which appeared in *Erkenntnis* in 1935, Neurath turned the charge of 'pseudorationalism' against Karl Popper's *Logik der Forschung*. This book, considered to be one of the major works of the philosophy of science of our time, had been published in the autumn of 1934 in the series *Schriften zur wissenschaftlichen Weltauffassung* edited by Schlick and Frank,[1] and had received sympathetic treatment from Carnap, who defended it in *Erkenntnis* against the criticisms of Reichenbach.[2]

A central chapter of the book was devoted to the 'empirical basis' of scientific knowledge, and one section of this chapter to the criticism of the 'psychologist position' advanced by Neurath and Carnap in their theory of 'protocol statements'.[3] Popper – who wrongly saw Neurath, instead of Carnap, as originator of the protocol theory[4] – believed that Neurath, in common with Carnap and all other exponents of logical positivism, made 'perceptual experience' the 'source of knowledge' (*die Erkenntnisquelle*).[5] Such an idea, he argued, was no different from the traditional sense-based and positivistic epistemological doctrine which established the certainty of scientific knowledge on the psychological datum of the immediacy of perceptual knowledge.[6]

This doctrine Popper rejected on the grounds, first, of the serious confusion he saw in it between psychology and the logic of cognition and, second, because of the indissoluble links it had with an inductivist conception of the logic of scientific discovery.[7] Neurath's sole advance beyond this traditional doctrine lay, he thought, in the recognition that the protocol statements, as records of perceptions, were not 'inviolable'.[8] But, although this 'step forward' represented a notable advance in the right direction, it led nowhere if it was not to be followed by another step: "We need a set of rules to limit the arbitrariness of 'deleting' (or else 'accepting') a protocol statement".[9]

In his view, however, Neurath failed to provide any rules of this kind, and was not in a position to provide any.[10] Without knowing it, he had 'thrown empiricism overboard', because his failure to adopt a rigorous methodological criterion meant that empirical sentences could no longer be distinguished from any other sort of proposition. If we could delete at will an observation sentence felt to be 'inconvenient', not only was it possible for us to rescue, in the manner of conventionalism, any system contradicted by experience, but also any arbitrary system could set itself up as 'empirical science'.[11] In Neurath's scheme the protocol statements had no precise part to play; they were "merely a relict – a surviving memorial of the traditional view that empirical science starts from perception".[12]

These criticisms contain some undoubtedly odd features. To begin with, Popper turns against Neurath the very charge of 'psychologism' (i.e. immediate perceptual experience taken as fount of cognitive certainty) which, as we have already found, Neurath had himself previously formulated in his clashes first with Wittgenstein and Schlick, and subsequently with Carnap. Secondly Popper repeats against Neurath the (opposite) criticism of relativism and methodological laxity[13] which, with little justification, Schlick had directed in 'Das Fundament der Erkenntnis' against Neurath, and then by extension against Popper himself,[14] intending that it should form his reply to the charge of 'psychologism' he had received from Neurath.

The incompatibility of these two aspects of Popper's attack on Neurath's protocol theory is manifest. And yet at no point in the course of his sharp polemic against Neurath's 'positivism' did he become at all aware of it.[15] It would appear that his aim of exalting the antipositivist importance and originality of his own *Basissätze* theory led him on the one side to yield to the temptation of playing down the critical aspects of Neurath's examination of the Vienna Circle's orthodoxy,[16] and on the other to attempt to disqualify as inconclusive the conventionalist solution Neurath had advanced with respect not only to observation sentences but also, more consistently than Popper, to any theory operating on a higher level .

2. NEURATH'S REPLY: *PROTOKOLLSÄTZE* AND *BASISSÄTZE*

Neurath was justified in replying that Popper was simply wrong (*irrt*) in asserting that he had conceived the theory of protocols as a 'psychologist' theory of the empirical basis. For the truth was that he had never seen the protocols as 'basic propositions' (*Elementarsätze*) belonging to an original or pure 'experiential' or 'phenomenal language' (*Erlebnissprache*, *phänomenale Sprache*), and as being capable of immediate correspondence with reality.[17] His very purpose in criticising and reformulating Carnap's theory of the protocols had been, quite simply, to raise "a protest against the theory of elementary propositions".[18]

Popper's doctrine of the *Basissätze* was in any case, he pointed out, comparable in many ways with his own theory of protocols.[19] Popper's 'basic propositions' were 'single' assertions accepted, as a result of agreements and decisions, by observers conveniently situated in time and space. The content of these assertions lay in the declaration that "an observable event is occurring in a certain individual region of space and time". [20] Put this way, the only difference between the two theories consisted in the fact that, whereas the protocols positively associated an observation account with the individual through whom it was transmitted (without, however, this giving them any connotation of immediacy or absoluteness), Popper believed that a non-psychologistic resolution of the problem of the relation between linguistic statements and experience would involve the terms 'observable' and 'observability' being restricted to the level of attributes of an observed physical event, while leaving their meaning undefined.[21]

But this was not a solution which Neurath found at all adequate. First, it made (realistic) reference to objective properties of physical bodies on a macroscopic level, and second, definition of the criteria of 'observability' of objects might prove very hard in disciplines other than physics, such as sociology and psychology. Still more significantly, Popper had ignored the relationship between a language and the individual users of it. This had led to his failure to recognise the historical variability and pluralism of scientific languages and to his supposition of the existence of a universal, semantically pure, scientific language.[22] With a reference to K. Ajdukiewicz, whose pragmatic conception of language and meaning was very close to the radical conventionalist theses of Le Roy,[23] Neurath raised the problems of social relativity, the historical variability in all languages of semantic values, and the difficulties of 'translation'. These were themes which were to be central to the self-critical thought of Wittgenstein's *Philosophische Bemerkungen* and *Philosophische Untersuchungen*,[24] and which, from Quine to Feyerabend and Kuhn, were to become the *locus classicus* of the English and American 'revolt against positivism' in the Sixties.[25]

"If", Neurath argued, "a primitive says 'the river runs through the valley', he uses a different definition of these terms from that of a European using the same phrase at a later date".[26] For Neurath, as for late Wittgenstein,[27] such semantic variability was the source of insuppressible ambiguity and imprecision in language. The imprecision of all terms formed "part of the nature itself of language",[28] and included also the propositions used in theoretical physics and chemistry, however great an impression of purity they gave.[29] At the same time it was the very imprecision, openness, and flexibility of language which made the historical continuity of linguistic communication possible and which put "people" in touch with "people, era with era, and scientist with scientist".[30]

3. TWO FORMS OF CONVENTIONALISM IN CONFLICT

In the light, however, of a historical approach to the development of science *(ein Modell der Forschungsgeschichte)* the central issue was not that of the semantic variability of the protocol statements or basic propositions.[31] Rather the crucial point was this: once it was seen that observation sentences did not form the source of scientific certainty, that they were neither original or unalterable, then it was necessary to go further and recognise that it was not possible to establish any generally outlinable method (*generell skizzierbare Methode*) of fixing logical criteria for acceptance or rejection of scientific terms and propositions, at whatever level they operated:

> we deny that the encyclopedia preferred by a scientist can be logically distinguished from others by using a method which can be generally outlined.[32]

It was at this point, the step immediately following recognition of the conventional nature of the observation sentences, that Neurath was furthest from agreement with Popper's *Logik.* Neurath's consistent conventionalism stood in opposition to Popper's attempt to combine the conventionalist solution of the problem of the empirical basis with both a strict deductivism in the testing of theories and a general realistic epistemology openly opposed to conventionalism.[33]

The step which Popper held to be necessary – and which, as Schlick had already done, he was to reproach Neurath for not having been able to take – was the establishment of a logical and methodological scheme for fixing universal criteria of procedural rationality in science. For Popper, this method lay in the (non-inductive) procedure through conjectures and refutations and in the experimental falsification of conjectural hypotheses in accordance with the *modus tollens* of classical logic. [34]

By means of highly detailed arguments Neurath advanced his claim that the falsificationist method could not be accepted as the universal standard of scientific rationality. Above all he argued that this was a function which no method could fulfil, for "at all times the whole apparatus of scientific propositions and all methods may be open to question".[35] The claim of Popper's falsificationism to present itself as 'the methodological paradigm of all empirical sciences' (*als das Paradigma aller Realwissenschaften*) was no more than a new incarnation of 'pseudorationalism': it denied the 'plurivocality' of the empirical sciences, and was an attempt, analogous to that of Laplace, to enclose them within the logic of the 'closed and perfect system' – nothing else, that is, than a new fitting-out of the 'old philosophical absolutism'[36] which had stood in the way of effective understanding of the praxis (*Forschungspraxis*) and history (*Forschungsgeschichte*) of scientific research.[37] In place of the 'absolutism of verification', Popper's philosophy quite simply substituted 'absolutism of falsification'. Thus falsificationism became no more acceptable than verificationism, since it led only to the same 'logical absolutisation' of scientific method.[38]

Popper's philosophy was guilty also of proposing a new 'ideal method', towards which, even though it was likely never to reach it exactly, science should aim – i. e. the construction of some system of empirical knowledge which would be in a position to use theoretical means to describe (*beschreiben*) 'our world', the 'world of real experience'.[39] But such an 'ideal' merely reinstated the traditional metaphysical and ontological aspiration towards a system of 'true' knowledge corresponding strictly, definitively, and univocally to the 'real world' (*wirkliche Welt*).[40]

Neurath then proceeded to develop a series of specific objections to Popper's position. These objections, prompt and timely as they were, constituted the first critique of falsificationism and anticipated by at least 30 years the reasoning behind the neo-Popperian revision of falsificationism to be seen in such writers as Lakatos, Agassi, or Watkins.[41] In addition these objections succeeded in outlining the role of a pragmatic and historical/sociological conception of science as an alternative to fundamentalism and justificationism. In this way Neurath anticipated, and even to some extent went beyond, the themes so typical of the English and American 'revolt against positivism'.[42]

Popper maintained that induction was a 'unfounded anticipation' which could not be reconstructed logically or systematically. At the same time, however, he attempted to invest falsificationism with the characteristics of a rigorous logical procedure. [43] He saw scientific hypotheses, in the form of universal laws or 'laws of nature', as not being able to be inferred or confirmed on the basis of a finite number of single observation sentences. On the contrary they could be negated by even one observation statement which set the contradictory evidence of one or more 'crucial experiments' against the universality of the laws and so permitted the advancement of a 'falsifying hypothesis'.[44]

Neurath denied that in the actual practice of scientific enquiry there was a methodological asymmetry between the experimental procedures of confirmation (or 'corroboration') and the experimental procedures of 'falsification', an asymmetry which corresponded – and had the possibility of corresponding – to the asymmetry between *modus ponens* and *modus tollens*. For him scientific hypotheses were not susceptible to induction, or to verification, or to 'falsification', if these terms implied the possibility of applying to the hypotheses formal rules which would operate as 'automatic' criteria of inference or testing.[45] Reasoned analysis led above all to the conclusion that one could not speak of 'inductive logic', because induction of theoretical hypotheses starting from observation sentences – as he had maintained since his work on physicalism in 1931 – had always to imply the adoption of methodological decisions (*die 'Induktion' beruht auf 'Entschluß'*)[46] and could not therefore admit of the rigorous inferences attained solely through logical deduction.[47]

And then it was necessary to realise, as Popper himself did, that repetition of observations and experiments and the systematic accumulation of empirical material did not add up, in Baconian fashion, to 'the method' by which (using inductive extrapolation) the formulation of scientific hypotheses was arrived at. Instead induction was a useful and frequently necessary practical means which acted in a great number of scientific disciplines as the premise of interpretation and theoretical elaboration.[48] Consequently Popper had been wrong in ascribing to Neurath an *a priori* 'belief' in the 'inductive method' and in thus assimilating his position to the views of Reichenbach (which were later taken up by Carnap). On the contrary these were views which he had explicitly criticised, both when he expressed his doubts about the attempt made by Reichenbach and Carnap to construct a 'scale of probabilities' applicable to theories and when he appealed to Mach's idea of the *Ökonomieprinzip* as a practical rule for limiting the number of empirical tests without starting from the metaphysical assumption of the 'constancy' *(Konstanz)* of natural phenomena.[49]

But, he continued, if it was not possible to discover in induction a general method for the formulation of empirical theories, then it was equally impossible to define a general method for their testing, still less to identify it, as Popper proposed, with experimental 'falsification'.[50] Popper was wrong to maintain that the scientist "should immediately discard a theory because of some negative results".[51] Rather he should endeavour to discover by means of a complex series of considerations whether the general theoretical structure, into whose framework the hypothesis not confirmed by experiment was to be fitted, was still capable of producing good results – and hence valid predictions – for the future. Rather than 'falsifying' a theory, negative experimental results could 'shake' a researcher's 'confidence' (*das Vertrauen erschüttern*) in a large-scale theoretical structure, but not automatically dictate its abandonment.[52] A successful researcher might simply decide to put on one side (*beiseite schieben*) what Popper called a 'falsifying hypothesis', for the reason that important general considerations allowed him to maintain that his acceptance of it would constitute a hinderance to the development of science. He could quite reasonably maintain that scientific development itself would put him in a position either to resolve the difficulty (*Einwand*) or to remove it.[53]

Popper's method demanded that a theory of wide scope should be considered falsified on the basis of negative results of a partial or marginal character, but this was a decision likely hardly ever to be taken by scientists engaged in practical work; for them positive results were of equal if not more importance than negative ones, a direct contradiction of the asymmetry of the falsificationist scheme.[54] Convincing examples could be seen in the history of science, such as the theory of the elementary particles of electricity which continued to be held despite the imposing mass of observations in apparent contradiction of it, and where the proofs of the theory were in fact considered more interesting than the 'disproofs'.[55] This was also the case – as Neurath was to add in his *Foundations of the Social Sciences* – with Newton's theory of gravity, which was constantly employed despite the fact that for some 150 years scientists continued to discover new elements of uncertainty or of contradiction in it; so useful was it in the analysis of the motion of bodies that few scientists took the trouble to point out its theoretical failings.[56]

In all such cases, had the methodological procedure of falsicationism been followed, theories of great consequence would never have been recognised, or, if recognised, would have been prematurely discarded.[57] The actual consequence of Popper's method seemed to be that, faced with a choice between an established general theory and new hypotheses built on experimental data which contradicted that theory, we should automatically favour the new hypotheses. And this, as Neurath was shortly later to observe when using the teaching of Duhem and Poincaré against Popper,[58] asserted a logically inconsistent methodological rule. When a theory encountered difficulties because of negative experimental results, it was in fact often logically possible, even at times reasonable and productive, to preserve the theory "by means of the introduction of a new variant".[59]

Still more serious was Popper's failure to notice that in such cases the decisive factor was the action taken by the "scientists engaged in practical research" (*Männer der Forschungspraxis*). Instead of merely reviewing, as Popper had, the possible objections of classical conventionalism (which would be valid at most for a discussion between traditional philosophers or pure theorists), what was necessary was a historical survey (*an Hand der Geschichte der Realwissenschaften*) of the kind of 'defensive actions' (*Abwehrbewegungen*) which 'practical scientists' actually took under such circumstances. For they were the people who experienced the effects of any eventual change in a theoretical paradigm.[60] It was always the case that 'practical scientists' (*die Praktiker der Forschung*), i.e. the effective practitioners of science, tended not to give importance to the deficiencies and contradictions present in a general theory, so long as that theory could be utilised without serious inconvenience and so long as 'no other more attractive hypothesis' presented itself.[61]

To Neurath's mind, therefore, theoretical as well as historical and sociological reasons demanded that the rigid Popperian model should be exchanged for a considerably more flexible scheme, which would view 'verification' and 'falsification' as at most no more than simple concept-limits (*Grenzbegriffe*),[62] which would place 'corroboration' (*Bewährung*) and 'weakening' (*Erschütterung*) of theoretical hypotheses on a level of absolute methodological symmetry, one beside the other.[63] A theory would be considered 'corroborated' (*bewährt*) when it was supported by a large number of favourable results and 'weakened' or 'shaken' (*erschüttert*) when one or more negative results conspired to shake (*erschüttern*) the confidence placed in it by scientists. The most accepted and stable theories would be those supported by a large number of positive results and weakened by a small number of negative results or by negative results of little practical relevance. In this sense, then, it would be permissible to speak of theories more plausible (*plausibler*) – but not more 'probable' – than others, although this could not be taken to mean that one could ever construct any kind of 'scale of plausibility' or define 'levels of plausibility' in theories, as Carnap had claimed in *Testability and Meaning* on the basis of his notion of 'levels of confirmation'.[64]

Retention or rejection of theories would remain a matter of decision for the scientific communities, and the decisions of such communities would be made on the basis of informal criteria varying from case to case.[65]

4. 'LAWS OF NATURE' AND EXISTENTIAL PROPOSITIONS: A CRITICISM OF THE CAUSALIST AND DEDUCTIVE MODEL OF SCIENTIFIC EXPLANATION

One corollary arising from the anti-inductivism and epistemological 'realism' of Popper's *Logik* was the adoption of a deductive model of scientific explanation in its most strict causalist form.[66] A second, closely connected with the first, was to take the falsifiability of theories by experiment as a criterion of those theories' scientific quality and therefore as a general line of demarcation between science and metaphysics.[67] A third was to see the *experimentum crucis* as a turning-point in the history of the development of science, a development which Popper saw as leading through successive approximations towards theories possessed of ever more elevated levels of universality.[68]

To each of these three points Neurath advanced specific objections, arguing first that scientific explanation required no absolute reference to 'laws of nature' of a strictly universal character; second that the methodological code developed by Popper offered no useful criterion for demarcation between science and metaphysics; and third that the historical view of the development of scientific theories in terms of successive approximations had no merit in it.

Neurath's critique is best understood in the light of the underlying tenets of Popper's falsificationism. Popper principally held that inductive generalisation had no methodological relevance with respect either to the 'context of discovery' or to the logical justification of theories. In his view no inductive inference existed to allow transition from observation sentences, which are singular, to universal statements of the 'laws of nature'.[69] Such a transition (induction through enumeration) was, he held – revealing the epistemological ontologism of his *Logik* – precluded for the reason that he saw 'laws of nature' not as simple 'numerical universals' but as strictly universal statements pertinent to an unlimited number of possible cases.[70] This was why he did not merely see existential propositions of the type 'some A are B' [71] as 'non-empirical' and metaphysical on the grounds that they were not, as he saw them, falsifiable on the basis of a finite number of contrary pieces of evidence; nor did he merely see probabilistic propositions as equally metaphysical and 'devoid of empirical meaning' on the grounds that they were neither verifiable nor falsifiable.[72] He required:

1. that all scientific laws should be constructed in accordance with the

logical form of universal quantification in sentences of the type 'all A are B';
2. that, whatever their origin with regard to 'context of discovery', the laws should refer not to a 'finite class of specific elements within a particular finite spatio-temporal region'[73] but to an 'unlimited number of individual elements';
3. that the laws should always start from the assumption that the world was not limited in time or space.[74]

In his view the concern of the scientist and especially of the physicist could not be reduced to the practical or technological matter of deducing predictions. The supreme task of the scientist, who could not be content with simple empirical generalisations, was to search for a causal explanation of phenomena, to discover *true* explanatory theories which described certain structural properties of the world on the basis of laws valid for a number of occasions unlimited in space or time.[75]

Such were the realistic and causalistic metaphysics from which Popper's *Logik* claimed to draw inspiration and towards which it offered encouragement. They stood in neat contrast, and sometimes open conflict, with the indeterministic and probabilistic tendencies present in theoretical physics in the Twenties and Thirties as a result mainly of the interpretations of quantum mechanics offered by Heisenberg and Bohr.[76] Starting from this conservative scheme, Popper advanced the thesis that scientific theories, whether explanatory or predictive, were *exclusively* susceptible to falsification: since they contained 'strictly universal' statements, they were not verifiable on the basis of a finite number of confirmations by experiment. Had Popper instead made reference to general laws having validity within a limited area and therefore universal only in a logical or numerical sense – as in 'all the inhabitants of the planet' or 'all the male citizens of this city' – he would obviously have been able to have affirmed their falsifiability even on the basis of a single negative instance, but he would not have been able to exclude verifiability from them. Numerical universals would have been able to be confirmed by a finite number of positive instances coinciding with the sum of cases predicted by the law – a high number in many cases, but not unlimited all the same. In this way induction through enumeration would have presented itself as legitimate.

These were also the reasons why Popper, in addition to constructing a nomological and deductive model of scientific explanation and prediction, also attributed the strictest causalist meaning to it.[77] According to him, every scientific explanation or prediction should always contain within the *explicans* from which the *explicandum* would be logically deduced, and alongside the singular propositions which defined the 'initial conditions' of the phenomena to be explained or predicted, one or more hypotheses with the character of 'universal laws of nature' sufficient to explain why one event rather than another had occurred (or to predict that it would occur).[78]

Neurath declared totally unacceptable the dogmatic imposition of this deductive and causalistic model of scientific explanation and prediction. This he did above all in the name of a general principle of methodological tolerance aiming to take account of the plurality of methods actually practised within the scientific disciplines.[79] From the standpoint of a 'tolerant empiricism' (*toleranter Empirismus*) it was not necessary for the so-called 'laws of nature' to have the character of strict universality.[80] There was in fact no theoretical or practical reason to prevent the scientist who intended to work with special caution (*aus besonderer Vorsicht*) from taking all the laws he used as valid within a limited sphere (*ein beschränktes Gebiet*) and from considering the universe itself as a finite entity.[81] Furthermore it was very doubtful – here again Neurath referred to the teaching of Mach – that science aimed, or ought to aim, to explain infinite classes of occurrences or that it had the function of extending the validity of the so-called 'laws of nature' without limit.[82]

Secondly, while not subjecting the 'deductive-nomological' model of explanation to direct criticism, Neurath objected to the historical variability not only of theories but also of the methods of their testing.[83] In the case of statistical explanation, for example, – a case, however, which Popper did not even take into account because, as we have seen, he treated probabilistic propositions as metaphysical and 'devoid of empirical meaning'[84] – the criteria of testing could vary in relation to the classes of reference preselected as the basis for statistical treatment.[85] More generally, as Duhem and Poincaré had shown and as the fate of many important scientific predictions attested, the absolute foundation did not exist to allow us to talk of a 'constancy of the rules for testing' (*eine Konstanz der Kontrollregeln*), even within closely defined periods in the history of science.[86]

As for the exclusion of 'existential propositions' from the realm of science on the grounds that, not being falsifiable, they should be considered as 'devoid of empirical content' and metaphysical, Neurath argued that Popper had forgotten that they played a part in the normal *praxis* of research and were useful in numerous scientific disciplines, such as astronomy, geology, and sociology, not to exclude areas of physics such as optics and acoustics, even if less often in these.[87] Suppose an astronomer were to predict that a comet would appear in a certain part of the sky, but without stating exactly when it would appear, there was no reason, he said, to consider this prediction metaphysical, even though it was obviously not falsifiable. If accurate and systematic observation of that part of the sky led to the comet actually being seen, not only could it be said that the hypothesis was confirmed and that it had empirical meaning, but also the astronomer would eventually be able to provide confirmation for a new and perhaps – to put it in Popperian fashion – 'very daring' (*sehr küne*) theory.[88]

It was therefore possible to develop a theory of scientific research which granted a 'legitimate function' to existential propositions, even though they were not falsifiable.[89] The criterion of falsifiability could not therefore be taken as a line of demarcation between scientific and metaphysical propositions. That criterion could not even be usefully applied to theoretical models of wide scope. Because these were not susceptible to direct experimental testing, Popper had ended by relegating them to the 'regions of metaphysics', as in the case of the ancient corpuscular theory of light.[90] Against this Neurath argued that there were no real reasons to exclude the scientific character of a hypothetical model which, albeit vaguely and without admitting of proper testing, showed that certain correlations between phenomena of light, at first not contained within a unitary explanation, could be derived from certain general premises in support of the hypothesis of a corpuscular theory. In other words, the scientific – or conversely metaphysical – quality of a general theoretical model could not be fixed in terms of a cut and dried theoretical division (*Schnitt*) between experimental falsifiability and non-falsifiability. The testing of theories, just like their formulation, was a matter of degree.[91] All that could be assessed was the greater or lesser practical reliability of the theoretical models whose experimental implications had been in diverse measure corroborated (*bewährt*) or weakened (*erschüttert*) by experience.[92]

For Neurath, then, Popper's formal criterion failed to establish "a clear line of demarcation between science and metaphysical ideas".[93] In the end it was at once too strict from the logical point of view and too vague (i.e. of uncertain application) from the empirical point of view.

5. *EXPERIMENTA CRUCIS*: AGAINST POPPER'S CONCEPTION OF SCIENCE AS AN ASYMPTOTIC PATH TOWARDS TRUTH

Somewhat surprisingly we find the 'positivist' Neurath going so far in his critique as to reprove the anti-inductivist Popper for an excessive propensity towards Baconian experimentalism.

In a central passage of his *Logik* Popper had argued that

> any empirical observation in science can be presented (by the description of experimental elements etc.) in such a way that anyone who has learned the appropriate technique would be able to test it.[94]

To Neurath's mind, this claim exaggerated beyond all measure the strength of experimentation and the ability to reproduce phenomena in the *praxis* of the empirical sciences and in the actual course of their historical development.[95] He found it significant that, 'notwithstanding all Duhem's warnings', Popper referred so often to the *experimentum crucis* as if to something really decisive (*endgültig*) when a choice was to be made between competing theories.[96] Popper was not aware that in many empirical disciplines, such as astronomy, geology and sociology, the experiment did not play the central part he imagined it did.[97] There were many cases of experimental results being solidly grounded in theoretical conception but without the capacity to be repeated in the laboratory and still, for all that, being held in the highest regard by scientists.[98] The history of the theory of relativity was a good example: here Mach, "with penetrating use of logical analysis and not on the basis of experiments",[99] had refuted Newton's theory of absolute space and had reinstated the problem of inertia, a major contribution to the results later achieved by Einstein.

Experimental results were not, in any case, ever to be viewed as 'crucial', as if the path to 'better' theories revealed itself historically as a process of successive approximations in which the *experimentum crucis* played the part, in typically Baconian fashion, of pointer in the right direction through the systematic elimination of false trails and blind alleys.[100] At several points in his *Logik* this figure was used by Popper to represent the development of science:

> it is always the experiment which saves us from following a track that leads nowhere: which helps us out of the rut, and which challenges us to find a new way;[101]

the falsifying experiment

contributes in its own peculiar way to the historical development of science as a process of step by step approximations;[102]

in the evolution of physics especially

one may discern something like a general direction – a direction from theories of a lower level of universality to theories of a higher level[103] [...] For a theory which has been well corroborated can only be superseded by one of a higher level of universality; that is, by a theory which is better testable and which, in addition, *contains* the old, well corroborated theory – or at least a good approximation to it.[104]

Neurath was vigorous in opposition to this experimentalist and cumulative view of the development of science. The only thing which was important in the historical succession of theories was, he argued, "the actual stock of the valid predictions". If

a theory T1 provides a group of successful predictions termed A and the theory T2 a group of successful predictions termed A + B, then it is possible to say that theory T2 is more successful (*erfolgreicher*) than theory T1 and that the stock of the predictions A is an approximation to the stock of predictions A + B. However, this in no way implies that the principles of theory T1 have to be an approximation to the principles of the more successful theory T2.[105]

This type of approximation, which was evidently logically false, hardly ever occurred even on the historical level. The most persuasive proof of this had been provided by Duhem when he showed "how little the different stages of gravitation theory could be seen as 'approximations' to the successive stages".[106]

Popper may have said that "science is not a system of certain, or well-established, propositions, nor is it a system which steadly advances towards a state of finality",[107] but Neurath observed that this declaration was thoroughly contradicted by his incremental theory of the development of science as a succession of ever more general theories and by his open profession of faith in the "regularities of the world which we can unveil and discover".[108] And again, in his use of metaphysical terminology, Popper revealed his 'pseudorationalistic' tendency to conceive science as knowledge of the 'real world'[109] and to interpret "the history of science as an asymptotic progression towards the 'Truth'".[110]

A 'rationalistic' conception of empiricism could only refute such an 'metaphysical illusion'[111] in the same way as it refuted any 'magical' conception of logic, which held logical analysis to be "a magical sieve which in some way gives us an automatic sifting of propositions".[112]

'Pseudorationalism' claimed to fix logically, by means of a general method, the limits of what it was absolutely still permissible to seek scientifically, while

> scrupulous research can do no more than provide, after penetrating analysis, a proper understanding of the decisions necessary to be taken to fix the limits of future work. It is precisely these decisions which leave their specific mark on each particular period of the history of science.[113]

Despite the fact that empirical rationalism held that decisions taken in science should be based exclusively on considerations of a scientific nature, and despite the fact that it refused to "equate somehow the social success of a doctrine with its logical validity",[114] its general pragmatic orientation *(attitude pragmatique)* compelled it

> to make room on a large scale for decisions within science and to affirm in particular that researchers are, in every aspect of their work, always dependent on their social situation.[115]

The opposition between a pragmatistic and historical/sociological conception of the methods and discoveries of science and an ontological and logicistic conception could not have been more marked. Popper's metaphor, by which he described the 'daring structures' of science as being raised, not on solid rock, but on a marsh – an edifice built on piles – is well known.[116] But in so far as this metaphor aimed to express the view-point of a fallibilist and non-dogmatic epistemology, opposed to the absolutism of the positivistic position, it ought not to be forgotten that science was for Popper built of daring structures resting on supports which, even if they stood on unfirm ground, were still of sufficient strength to sustain the structures. For Neurath science was not an edifice built on a marsh but a ship in open sea, continually forced to carry out works of self-repair and restructuring and to struggle against 'heavy gales and thundering waves'. The theoretical – not simply predictive and practical – ways of science were not linear and progressive, but uncertain and reflexive. Human knowledge was the endless attempt to construct a ship from the remains of preceding wrecks.[117]

NOTES

[1] In his 'Autobiography' (in P.A. Schilpp, ed., *The Philosophy of Karl Popper*, p. 67) Popper refers to the complex circumstances of the publication of his book; see also R.

Carnap, 'Intellectual Autobiography', in P.A. Schilpp (ed.), *The Philosophy of Rudolf Carnap*, pp. 30–1. Neurath too, like other members of the Vienna Circle, had known Popper's complete typescript since 1932.

[2] Cf. Carnap's review in *Erkenntnis, 5* (1935), pp. 290–4; see in addition Reichenbach's critical essay 'Über Induktion und Wahrscheinlichkeit. Bemerkungen zu Karl Poppers 'Logik der Forschung'', *Erkenntnis, 5* (1935), pp. 267–84, English trans. in H. Reichenbach, *Selected Writings 1909–1953*, ed. by M. Reichenbach and R.S. Cohen, vol. 2, Dordrecht and Boston: D. Reidel, 1978.

[3] Cf. K. R. Popper, *Logik der Forschung*, pp. 53–5 (*Über die sogenannten 'Protokollsätze'*), English trans. pp. 95–7. See also V. Kraft, 'Popper and the Vienna Circle', in P.A. Schilpp (ed.), *The Philosophy of Karl Popper*, pp. 185–204.

[4] Cf. K. R. Popper, *Logik der Forschung*, p. 232, note 1, English trans. p. 95, note 1. Carnap used for the first time the term *Protokollsätze* in his essay 'Die Physikalische Sprache als Universalsprache der Wissenschaft', *Erkenntnis, 2* (1931), pp. 432–65.

[5] Cf. K.R. Popper, *Logik der Forschung*, p. 52, English trans. p. 94.

[6] Ibid., p. 54, English trans. p. 96.

[7] Ibid., p. 51, English trans. p. 93. This is why I would not agree with Pera's opinion that "as to the contention about protocols, which divided the Vienna Circle, Popper took sides with Neurath and Carnap" (M. Pera, *Popper e la scienza su palafitte*, Roma and Bari: Laterza, 1981, p. 198, note 14).

[8] K. R. Popper, *Logik der Forschung*, pp. 54–5, English trans. pp. 96–7.

[9] Ibid., loc. cit.

[10] Ibid., loc. cit.

[11] Ibid., p. 54, English trans. p. 97. In 'Falsification and the Methodology of Scientific Research Programmes', in I. Lakatos and A. Musgrave (eds.), *Criticism and the Growth of Knowledge*, pp. 113–4, 156, Lakatos hastily subscribes to Popper's evaluations. See also, in a similar vein, G. Andersson, 'How to Accept Fallible Test Statements? Popper's Criticist Solution', pp. 52–65. For a precise reconstruction of Neurath's criticism of Popper see rather G. Giorello, 'Il falsificazionismo di Popper', in L. Geymonat (ed.), *Storia del pensiero filosofico e scientifico*, vol. VII, pp. 466–74; see also G. Giorello, 'Introduzione' to I. Lakatos and A. Musgrave (eds.), *Critica e crescita della conoscenza*, Milano: Feltrinelli, 1976, pp. 16–9.

[12] K.R. Popper, *Logik der Forschung*, p. 54, English trans. p. 97.

[13] 'Permissive methodology' is Lakatos' expression in 'Falsification and the Methodology of Scientific Research Programmes', p. 113.

[14] Cf. M. Schlick, 'Über das Fundament der Erkenntnis', *Erkenntnis, 4* (1934), p. 83.

[15] For the existence of a particular tension between Popper and Neurath see Popper's testimony in 'Memories of Otto Neurath', *ES*, pp. 55–6.

[16] Returning to the question of 'protocols' twenty years later (cf. P.A. Schilpp, ed., *The Philosophy of Rudolf Carnap*, pp. 199–200), Popper showed even more openly this kind of attitude. After mentioning a (supposed) pre-physicalistic and solipsistic stage of Neurath's philosophical development, Popper maintains that Neurath's conception of 'protocols' was only an effort to embed the old solipsistic starting point ('our own observational experiences') within physicalism. According to Popper, in his essay 'Über Protokollsätze' Carnap was not aware of this aspect; this is why he wrongly equated Neurath's conception of 'protocols' and his own (i. e. Popper's) theory of *Basisätze*. According to Popper, while Neurath still assumed 'our own

sense-experiences' as the foundation of scientific knowledge, his own (i.e. Popper's) theory of *Basissätze* held that the tester would stop only when he had arrived at a statement "about some easily and intersubjectively observable *behaviour of a physical body*" (p. 200); see also K.R. Popper, 'Replies to My Critics' in P.A. Schilpp (ed.), *The Philosophy of Karl Popper*, pp. 969–70. According to Popper's interpretation the difference between the two positions lay in the fact that he, Popper, "never believed in induction [...] whereas Neurath did believe in induction" (P.A. Schilpp, ed., *The Philosophy of Rudolf Carnap*, p. 200). Only afterwards, Popper maintains, did Carnap perceive this difference and converted himself to (Popper's) criticism of psychologism and inductivism. As to Popper's inclination to suppose that his own criticism was decisive for Carnap's intellectual development and for the whole Vienna Circle's philosophical evolution, see Carnap's annoyed reply: "I appreciate the influence of Popper's ideas, but I am not sure whether they played quite the central role in the development of my views which he ascribes to them" (ibid., p. 880).

[17] O. Neurath, 'Pseudo', [1935a], pp. 361–2, *PP*, pp. 128–9.

[18] Ibid., p. 362, *PP*, p. 128.

[19] Ibid., loc. cit., *PP*, p. 129.

[20] K.R. Popper, *Logik der Forschung*, p. 60, English trans. p. 103.

[21] O. Neurath, 'Pseudo', [1935a], pp. 362, 363, *PP*, pp. 129–30; cf. K.R. Popper, *Logik der Forschung*, p. 60, English trans. p. 103.

[22] O. Neurath, 'Pseudo', [1935a], pp. 362–3, *PP*, pp. 129–30.

[23] Cf.: K. Ajdukiewicz, 'Sprache und Sinn', *Erkenntnis, 4* (1934), pp. 137–8; K. Ajdukiewicz, 'Das Weltbild und die Begriffsapparatur', ibid., pp. 259–87.

[24] Cf. A. Gargani, *Introduzione a Wittgenstein*, pp. 43–79.

[25] See for instance: F. Suppe, *The Structure of Scientific Theories*, Urbana (Ill.): University of Illinois Press, 1977, pp. 62–232, 617–730; W.B. Weimer, *Notes on the Methodology of Scientific Research*, Hillsdale (N.J.): Lawrence Erlbaum Associates, 1979, pp. 8–71; R. Rorty, *Philosophy and the Mirror of Nature*, pp. 257–311; M. Hesse, *Revolutions and Reconstructions in the Philosophy of Science*, 1980, pp. vii-xxvi.

[26] O. Neurath, 'Pseudo', [1935a], p. 363, *PP*, p. 129.

[27] Cf. A. Gargani, *Introduzione a Wittgenstein*, pp. 73–5 ff.

[28] "Die *Unbestimmtheit* aller Termini [...] gehört mit zum Wesen der Sprache" (O. Neurath, 'Einheit der Wissenschaft als Aufgabe', p. 20, [1935], *PP*, p. 118).

[29] O. Neurath, 'Pseudo', [1935a], p. 363, *PP*, p. 129.

[30] O. Neurath, 'Pseudo', [1935a], pp. 362–3, *PP*, p. 129; O. Neurath, 'Einheit der Wissenschaft als Aufgabe', [1935], pp. 20–1, *PP*, p. 118.

[31] O. Neurath, 'Pseudo', [1935a], p. 363, *PP*, p. 129.

[32] Ibid., p. 355 ("*Wir bestreiten, daß sich die von einem Forscher bevorzugte Enzyklopädie mit Hilfe einer generell skizzierbaren Methode logisch aussondern läßt*"), *PP*, p. 123.

[33] Cf.: K.R. Popper, *Logik der Forschung*, pp. 40–2, English trans. pp. 78–81; K.R. Popper, 'Three Views Concerning Human Knowledge', in K.R. Popper, *Conjectures and Refutations*, London: Routledge and Kegan Paul, 1963, pp. 97–119.

[34] The categories *modus ponens* and *modus tollens* were used in the logic of the seventeenth century to denote the two modes of hypothetical syllogism: 'if A then B, A is true, then B is true'; 'if A then B, B is false, then A is false'.

[35] O. Neurath, 'Pseudo', [1935a], p. 355, *PP*, p. 122.
[36] Ibid., pp. 355, 361, 365, *PP*, pp. 123, 128, 131. See also: O. Neurath, 'Einzelwissenschaften, Einheitswissenschaft, Pseudorationalismus', [1936a], *GpmS*, p. 708, *PP*, p. 137; O. Neurath, 'Unified Science as Encyclopedic Integration', in *International Encyclopedia of Unified Science*, vol. I, no. I, Chicago: The University of Chicago Press, 1938, pp. 20–1.
[37] O. Neurath, 'Pseudo', [1935a], p. 353, *PP*, p. 121.
[38] Ibid., pp. 355, 365, *PP*, pp. 123, 131. See also: O. Neurath, 'L'encyclopédie comme 'modèle'', *Revue de Syntèse*, XII, 2, p. 189, *PP*, p. 147; O. Neurath, 'Unified Science and Its Encyclopedia', [1937a], p. 276, *PP*, p. 180.
[39] K. Popper, *Logik der Forschung*, p. 11, English trans. p. 39.
[40] O. Neurath, 'Pseudo', [1935a], pp. 354, 364, *PP*, pp. 122, 130.
[41] See for instance: I. Lakatos, 'Falsification and the Methodology of Scientific Research Programmes'; J. Agassi, 'Towards an Historiography of Science', *History and Theory. Studies in the Philosophy of History*, *2* (1963); J.W.N. Watkins, 'Hume, Carnap and Popper', in I. Lakatos (ed.), *The Problem of Inductive Logic*, Amsterdam: North Holland, 1968, pp. 271–82.
[42] Cf.: H. Rutte, 'Der Philosoph Otto Neurath', in F. Stadler (ed.), *Arbeiterbildung in der Zwischenkriegszeit*, pp. 76–8; R. Haller, 'Das Neurath-Prinzip. Grundlagen und Folgerungen', ibid., pp. 86–7.
[43] O. Neurath, 'Pseudo', [1935a], p. 356, *PP*, p. 123.
[44] Ibid., p. 354, *PP*, p. 121.
[45] Ibid., p. 356, *PP*, p. 124; O. Neurath, 'Einzelwissenschaften, Einheitswissenschaft, Pseudorationalismus', [1936a], *GpmS*, p. 707, *PP*, p. 136.
[46] O. Neurath, 'Physikalismus', [1931d], *GpmS*, p. 418, *PP*, p. 53. See also O. Neurath, *Empirische Soziologie*, [1931e], *GpmS*, p. 517, *ES*, p. 407.
[47] O. Neurath, 'Zur Induktionsfrage', *Erkenntnis*, *5* (1935), pp. 173–4.
[48] O. Neurath, 'Pseudo', [1935a], p. 359, *PP*, pp. 126–7.
[49] Cf. Popper's polemical contribution to *The Philosophy of Rudolf Carnap*, ed. by P.A. Schilpp, p. 200. In his intellectual autobiography (in *The Philosophy of Karl Popper*, ed. by P.A. Schilpp) Popper declares that "Carnap's and Hempel's highly favourable reviews of *Logik der Forschung* were promising signs, and so, in another way, were attacks by Reichenbach and Neurath" (p. 71). Here Popper totally overlooks the fact that Neurath's critical arguments against falsificationism were from many points of view, particularly with respect to the problem of induction, quite opposite to the arguments advanced by Reichenbach. See: H. Reichenbach, 'Über Induktion und Wahrscheinlichkeit. Bemerkungen zu Karl Poppers 'Logik der Forschung'', pp. 267–84; H. Reichenbach, 'Zur Induktionsmaschine', *Erkenntnis*, *5* (1935), pp. 172–3; O. Neurath, 'Prognosen und Terminologie in Physik, Biologie, Soziologie', in *Travaux du IX Congrès international de philosophie*, IV, *L'unité de la science: la méthode et les méthodes*, Ie partie, *Actualités Scientifique et Industrielles*, no. 533, Paris: Hermann & Cie, 1937, now also in *GpmS*, pp. 789–90; O. Neurath. 'Universal Jargon and Terminology', [1941], p. 138, *PP*, p. 222; O. Neurath, *Foundations*, [1944a], p. 24 ("There are scholars who try to find certain rules for making inductions; but I can hardly imagine how these could be a substitute for decision"); O. Neurath, 'Prediction and Induction', *Analisi*, *1* (1946), no. 3, p. 1 ("Hume's renunciation of the induction and causality business did not satisfy research

workers, who wanted to have a kind of 'justification' for their everyday technique. I should like to call their attitude 'pseudorationalism'"), now also in *PP*, p. 243.
50 O. Neurath, 'Pseudo', [1935a], p. 355, *PP*, p. 123.
51 Ibid., p. 356, *PP*, p. 124.
52 Ibid., loc. cit.
53 Ibid., loc. cit.
54 Ibid., loc. cit.; O. Neurath, *Foundations*, [1944a], p. 26.
55 O. Neurath, 'Pseudo', [1935a], p. 359, *PP*, pp. 126–7 ("On the contrary Popper wishes to see forceful decisions forcefully founded. That is in fact a basic tendency of many pseudorationalist endeavours that should be explained perhaps with the help of the 'psychology of decision'" (ibid., pp. 359–60, *PP*, p. 127).
56 O. Neurath, *Foundations*, [1944a], p. 26.
57 O. Neurath, 'Pseudo', [1935a], pp. 356–7, *PP*, pp. 123–4. Pressed by Kuhn's criticism, Popper himself was to end up at the beginning of the Seventies by admitting that a certain measure of dogmatism was necessary: "I realized the opposite: the value of a *dogmatic* attitude: somebody had to defend a theory against criticism, or it would succumb too esasily, and before it had been able to make its contributions to the growth of science" (*Objective Knowledge*, Oxford: Oxford University Press, 1979, p. 30).
58 O. Neurath, 'Prognosen und Terminologie in Physics, Biologie, Soziologie', [1937b], *GpmS*, p. 793.
59 Ibid., loc. cit. ("Wenn wir auf Schwierigkeiten stoßen, die sich bei Überprüfung der Theorien ergeben, kann man durch Wahl einer anderen *Variante* unserer Auswahl die Theorie aufrechterhalten").
60 O. Neurath, 'Pseudo', [1935a], p. 357, *PP*, p. 124. The affinity of these theses with T. Kuhn's antipositivistic philosophy of science does seem surprising: cf. R. Hegselmann, 'Otto Neurath. Empiristischer Aufklärer und Sozialreformer', pp. 44–5.
61 O. Neurath, *Foundations*, [1944a], pp. 25–6.
62 O. Neurath, 'Einzelwissenschaften, Einheitswissenschaft, Pseudorationalismus', [1936a], *GpmS*, p. 707, *PP*, p. 136; O. Neurath, 'Physikalismus und Erkenntnisforschung', I, [1936f], *GpmS*, p. 752, *PP*, p. 161.
63 O. Neurath, 'Pseudo', [1935a], pp. 355–7, *PP*, pp. 123–4; O. Neurath, 'Einzelwissenschaften, Einheitswissenschaft, Pseudorationalismus', [1936a], *GpmS*, p. 707, *PP*, p. 136. Literally 'Erschütterung' means 'shake', 'shaking', or 'shock' and it is used by Neurath as a rather unelegant neologism coming from '*erschüttern das Vertrauen*' (to challenge, to weaken one's confidence).
64 O. Neurath, 'Universal Jargon and Terminology', [1941], p. 139, *PP*, p. 222 ("I cannot agree with Carnap when he adds to my proposal dealing with 'corroborating' and 'weakening' his proposal dealing with 'degrees' of confirmability"). Cf. R. Carnap, 'Testability and Meaning', *Philosophy of Science*, *3* (1936), no. 4, pp. 419–71, and *4* (1937), no. 1, pp. 1–40.
65 O. Neurath, 'Pseudo', [1935a], pp. 357, 360, *PP*, pp. 124, 127; O. Neurath, *Le développement du Cercle de Vienne et l'avenir de l'empirisme logique*, [1936], p. 55.
66 Cf. K.R. Popper, *Logik der Forschung*, pp. 26–40, English trans. pp. 59–77.
67 Ibid., pp. 14–15, 40–51, English trans. pp. 43, 78–92.
68 Ibid., pp. 181, 182, 206, 222, English trans. pp. 246, 247, 277, 301.
69 Ibid., pp. 28–9, English trans. pp. 62–4.

[70] Ibid., loc. cit.
[71] Ibid., p. 33, English trans. p. 69.
[72] Ibid., pp. 132, 139, English trans. pp. 190, 198.
[73] Ibid., pp. 28–9, English trans. pp. 62–3.
[74] Ibid., p. 28, English trans. p. 63.
[75] Ibid., pp. 27–8, English trans. pp. 51–2.
[76] Heisenberg and Bohr had a direct influence on some members of the Vienna Circle, particularly on Philipp Frank and Neurath. *Erkenntnis* published the paper presented by Heisenberg in Könisberg in 1930, and the paper presented by Bohr at the II International Congress for the Unity of Science held in Copenhagen in 1936. Cf.: W. Heisenberg, 'Kausalgesetz und Quantenmechanik', in *Erkenntnis, 2* (1931), pp. 172–82; N. Bohr, 'Kausalität und Komplementarität', ibid., *6* (1936), pp. 293–302. Bohr also contributed to *Foundations of the Unity of Science*, with a short article which appeared in the first booklet of the series (*Encyclopedia and Unified Science*, p. 28). Popper in *Logik der Forschung* (pp. 172–81, English trans. pp. 236–46) tried to criticise Bohr and Heisenberg's 'uncertainty principle' and ended up by making a serious mistake of 'imaginary experimentation', with which he was charged by Weizsäcker and Heisenberg himself. Without making a specific analysis of the complex problems of quantum mechanics, Neurath expressed his doubts about Popper's attempt to refute the indeterministic import of Heisenberg's principle of uncertainty (cf. O. Neurath, 'Pseudo', [1935a], p. 361, *PP*, p. 128).
[77] K.R. Popper, *Logik der Forschung*, pp. 26–8, English trans. pp. 59–62.
[78] Popper presented his model of nomological-deductive explanation in *Logik der Forschung*, pp. 26–8, English trans. pp. 59–60, and developed it in *The Open Society and Its Enemies*, London: George Routledge & Sons, Ltd., 1945, pp. 262–3, 362–4, and in 'The Aim of Science', *Ratio, 1* (1957–58), n. 1, pp. 24–35.
[79] O. Neurath, 'Pseudo', [1935a], p. 354–5, *PP*, pp. 122–3.
[80] O. Neurath, 'Prognosen und Terminologie in Physik, Biologie und Soziologie', [1937b], *GpmS*, pp. 791, 792.
[81] O. Neurath, 'Pseudo', [1935a], p. 357, *PP*, p. 125.
[82] O. Neurath, 'Prognosen und Terminologie in Physik, Biologie und Soziologie', [1937b], *GpmS*, p. 792. Neurath had already maintained this thesis in 'Wege der Wissenschaftlichen Weltauffassung', [1930/31], p. 119, *PP*, p. 42, and in the review of P. Frank's book *Das Kausalgesetz und seine Grenzen*. Cf. O. Neurath, 'Die 'Philosophie' im Kampf gegen den Fortschritt der Wissenschaft', *Der Kampf, 25* (1932), pp. 385–9, now in *GpmS*, pp. 571–6.
[83] O. Neurath, 'Prognosen und Terminologie in Physik, Biologie und Soziologie', [1937b], *GpmS*, pp. 792–3.
[84] K.R. Popper, *Logik der Forschung*, pp. 132, 138–9, English trans. pp. 190, 197–8.
[85] O. Neurath, 'Prognosen und Terminologie in Physik, Biologie und Soziologie', [1937b], *GpmS*, p. 793.
[86] Ibid., loc. cit.
[87] O. Neurath, 'Pseudo', [1935a], pp. 357–8, *PP*, p. 125.
[88] Ibid., p. 358, *PP*, p. 125.
[89] Ibid., pp. 357, 358, *PP*, p. 125.
[90] Ibid., p. 358, *PP*, p. 125. Cf. K.R. Popper, *Logik der Forschung*, p. 206, English trans. p. 278.

91 O. Neurath, 'Pseudo', [1935a], pp. 358, 360, *PP*, pp. 125–7.
92 Ibid., p. 359, *PP*, p. 125.
93 K. Popper, *Logik der Forschung*, p. 11, English trans. p. 39.
94 Ibid., p. 57, English trans. p. 99.
95 O. Neurath, 'Pseudo', [1935a], pp. 358–9, *PP*, pp. 126–7.
96 Ibid., p. 360 ("Aus dieser ganzen Haltung heraus erklärt sich wohl, weshalb Popper, trotz aller Warnungen Duhems, so gern vom 'experimentum crucis' spricht"), *PP*, p. 127.
97 Ibid., p. 358, *PP*, p. 125.
98 Ibid., p. 360, *PP*, p. 127.
99 O. Neurath, 'Einheitswissenschaft und Psychologie', [1933], *GpmS*, p. 598; O. Neurath, 'Der Logische Empirismus und der Wiener Kreis', German trans. from Swedish (*Theoria*, 2, 1936, no. 1), *GpmS*, pp. 740–1; O. Neurath, 'Unified Science and Its Encyclopedia', [1937a], p. 268, *PP*, p. 174; O. Neurath, 'Einheitswissenschaft als empiristische Synthese', German trans. from Dutch (*Synthese*, 3, 1938, no. 1), *GpmS*, p. 828.
100 In *The Logic of Scientific Discovery* (p. 78, note 1) Popper maintains, against Duhem, the possibility of falsifying *experimenta crucis*. See also the essay 'Three Views Concerning Human Knowledge', in K.R. Popper, *Conjectures and Refutations*, p. 112.
101 K.R. Popper, *Logik der Forschung*, p. 199, English trans. p. 268.
102 Ibid., loc. cit.
103 Ibid., p. 205, English trans. p. 276.
104 Ibid., loc. cit.
105 O. Neurath, 'Pseudo', [1935a], p. 363, *PP*, p. 130; O. Neurath, *Le développement du Cercle de Vienne et l'avenir de l'empirisme logique*, [1936], p. 55.
106 O. Neurath, 'Pseudo', [1935a], p. 363, *PP*, p.130. Popper's implicit reply can be found in 'The Aim of Science', pp. 29–35.
107 R. K. Popper, *Logik der Forschung*, p. 207, English trans. p. 278. Cf. O. Neurath, 'Pseudo', [1935a], p. 364, *PP*, p. 130.
108 K.R. Popper, *Logik der Forschung*, pp. 186–7, English trans. pp. 252–3.
109 O. Neurath, 'Pseudo', [1935a], p. 364, *PP*, p. 130. In his 'Replies to My Critics' (in P.A. Schilpp, ed., *The Philosophy of Karl Popper*, pp. 971, 1183) Popper writes: "Trying to look back to those days from a distance of almost forty years, it seems to me that only Neurath and Schlick took my attitude towards metaphysics seriously, and thus realized that I really was not a positivist and, in some sense of the word, not even an empiricist, but close to a position Neurath called 'rationalism' or 'pseudorationalism'". In a note he adds: "In his last section [of 'Pseudorationalismus der Falsifikation'] Neurath speaks quite correctly of my 'metaphysical tendencies' which make me 'friendlier towards Kant and other metaphysicists than towards the group of thinkers whom he … describes as *the* positivists'. Perhaps I may repeat here that I am a realist who makes the following concession to idealism: all our theories are our inventions, as Kant says; we try to impose them upon the world. Where Kant was wrong, under the influence of Newton's success, is that he did not see how often we err, and fail to impose theories on the world".
110 O. Neurath, *Le développement du Cercle de Vienne et l'avenir de l'empirisme logique*, [1936], p. 55. See in addition O. Neurath, 'Prediction and Induction',

[1946a], p. 1 ("There is no judge in a chair who decides who is nearer to the truth. There is no way of 'impartiality' or 'scientific objectivity'"), *PP*, p. 243.
[111] O. Neurath, *Le développement du Cercle de Vienne et l'avenir de l'empirisme logique*, [1936], p. 55.
[112] Ibid., loc. cit.
[113] Ibid., p. 56.
[114] Ibid., p. 57.
[115] Ibid., loc. cit.
[116] K.R. Popper, *Logik der Forschung*, pp. 66–7, English trans. p. 111.
[117] O. Neurath, *Foundations*, [1944a], p. 47. Cf. H. Blumenberg, *Schiffbruch mit Zuschauer. Paradigma einer Daseinsmetapher*, chap. 6, passim.

CHAPTER 5

THE UNITY OF SCIENCE AS A HISTORICO-SOCIOLOGICAL GOAL: FROM THE PRIMACY OF PHYSICS TO THE EPISTEMOLOGICAL PRIORITY OF SOCIOLOGY

1. FROM 'UNIFIED SCIENCE' TO THE ENCYCLOPEDIC 'ORCHESTRATION' OF SCIENTIFIC LANGUAGE

In September 1935 the first International Congress for the Unity of Science took place at the Sorbonne in Paris with a large attendance by European as well as by a number of American scientists. Neurath, the moving force behind the conference, was also its leading light.[1] It proved to be a triumph for mid-European logical neopositivism and put Neurath's plan for a new *'Encyclopedia'* of scientific knowledge on an international footing.[2] From this point until his death at Oxford in 1945, the majority of his own writings were devoted to the problem of the encyclopedic integration of science, while his wider social and editorial energies were absorbed by the practical plan for an *International Encyclopedia of Unified Science*. Despite the obstacles presented in turn by the advent of National Socialism and by the outbreak of war, Neurath's project materialised, in 1935, in the foundation at the Hague of the International Institute for the Unity of Science[3] and, from 1938, in the publication of *Foundations of the Unity of Science* under the auspices of the University of Chicago.[4]

In the first formulations of Neurath's programme physicalism and unity of science were closely connected. As time went on, however, further development of the theory of physicalism gave place to the predominantly practical demands of organising and publishing the *neue Enzyklopädie*, whose planning and composition eventually became his main concern. This involved recapitulation – and in some respects, as we shall see, weakening and even alteration – of the original tenets of physicalism.[5] The concept of 'unified science' (*Einheitswissenschaft*), strictly envisaged at first, took on more open, pragmatic and tolerant features. At the end appeal was made to the encyclopedic 'model' mainly for its polemical function against 'pseudorationalistic' conceptions of science as a 'closed' or ideal system. Even in the matter of terminology, Neurath came to use expressions such as 'empirical rationalism' or 'scientific empiricism' (a borrowing from Gregorius Itelson)[6] to denote his encyclopedic standpoint in preference to the former, more specific, 'logical empiricism' or 'physicalism'.[7]

Neurath's neo-encyclopedism referred to the *Encyclopédie* as its most immediate historical antecedent, reflecting in many ways its scientific optimism and spirit of tolerance. There was conscious reference also to the philosophical aspects of the ideal of the unification of science shown by d'Alembert and his precursors and successors.[8] At several points the *Pansophia* and *Orbis Pictus* of Comenius, Paul Otlet's *La cité mondiale*, projected encyclopedic systems such as Leibniz' *Atlas universalis* and *Characteristica universalis*, the schemes of Saint-Simon, Comte, Mill, Spencer, and Wundt were cited as the most important contributions in the history of western philosophy on the subject of the encyclopedic unification of knowledge.[9]

In contrast with this philosophical tradition – whose basic premises of value it shared – Neurath's encyclopedism was distinguished from the outset by the clear prevalence it gave to the linguistic element as 'grammatical' vehicle of encyclopedic unification.[10] Even in his early work unification of language had been seen as a historical and sociological aim (*ein historisches Faktum in soziologischem Sinne*),[11] to be achieved through integrated organisation of scientific work and not as the result of a kind of 'philosophical centralisation' of the sciences by means of a 'superscience' claiming, in Comtean manner, to provide normative regulation of the actual work of scientists.[12] In Neurath's view, unification of science required the active participation of scientists and the fostering amongst them of agreement about any given project.[13] The gradual adoption of a homogeneous linguistic apparatus would in turn form the unitary grammatical basis of the individual scientific disciplines, capable of liberating them from metaphysics and of making them the acknowledged instrument of cultural regeneration and the advancement of human life.

Demarcation between science and metaphysics was not therefore, as Neurath saw it, an end to be achieved, as Carnap held, via the logical analysis of language nor, as Popper claimed, through the application of a logically rigorous method of testing theories. Victory over metaphysics was conceived of as the historical outcome of a cultural battle within the world of science which would result above all in scientists being occupied in the collective development of an empiricist 'universal jargon'.[14] Actual practice of the empirical sciences was what would gradually purge ordinary speech of its metaphysical elements and create conditions favourable to the formation of a 'jargon' which would comprise both everyday language and the technical language of scientists.[15] Only collective work and critical examination on the part of the scientists could lead to the gradual abandonment of the traditional 'ontological' problems of theology and speculative metaphysics as being unimportant in the operative procedures of scientific prediction, and so result also in the elimination from the grammar of scientific discourse of propositions meaningful solely within the context of those problems.[16]

Such a universal process of acculturisation and scientific education would undoubtedly have led, in Neurath's physicalism of the early Thirties, to the exclusion from the grammar of the empirical sciences (including psychology, anthropology, and sociology) of terms and hypotheses not pertaining to spatio-temporal co-ordinates and consequently not directed towards the practical aim of prediction.[17] Hence arose the stress with which he located the paradigm of a non-metaphysical linguistic formula within the physicalist 'jargon' of daily life – rather than in the conceptual structures of physics – for the reason that it pertained to relations between physical objects determined in space and time.[18] Here too lay the origin of the inflexibility of his battle against 'mentalism' and 'vitalism' in psychology and sociology, a battle which appeared, at least to begin with, to be founded on an uncritical amalgam of the most unyielding precepts of Watsonian and Pavlovian behaviourism.[19]

However much the project for the unification of the language of science arose within the general bounds of physicalism, especially with respect to the linguistic distinction between science and metaphysics, it remains the case that it was free, even in its original version, of the dogmatism, epistemological realism, and utopianism with which it has so often been charged.[20] The element which commentators and critics – Popper, once again, more than any – have failed to appreciate is the essential disparity between Carnap's thesis of the logical unity of the language of science[21] and Neurath's project for an encyclopedic 'orchestration' of the discursive processes involved in scientific work.[22]

2. POPPER'S OBJECTIONS TO THE PROJECTS OF NEURATH AND CARNAP

The possibility of creating a single universal language, Popper maintained in 'The Demarcation between Science and Metaphysics'[23], had already been disproved – even before it had been advanced by Carnap in 1932 – by the second of Gödel's theorems of incompleteness. Subsequently Tarski had demonstrated in 1933 in 'Der Wahrheitsbegriff in den formalisierten Sprachen'[24] the inconsistency of any universal language.

In spite of all this – Popper wrote – the doctrine has survived [...] and the so-called 'International Encyclopedia of Unified Science', which was founded upon this doctrine (despite my opposition, at the 'First Congress for Scientific Philosophy'in Paris, 1935) is still being continued. It will remain a monument to a metaphysical doctrine, once passionately held by Neurath and brilliantly wielded by him as a major weapon in the anti-metaphysical crusade. For no doubt the strong philosophical belief which inspired this forceful and lovable person was, by his own standards, purely 'metaphysical'. A unified science in a unified language is really nonsense.[25]

And he adds in a note:

In Paris, I opposed the foundation of the *Encyclopedia*. (Neurath used to call me 'the official opposition' of the Circle, although I was never so fortunate as to belong to it). I pointed out, among other things, that it would have no similarity whatever to an encyclopedia as Neurath conceived it, and that it would turn out to be another series of *Erkenntnis* articles. At the Copenhagen Congress, in 1936, which Carnap did not attend, I tried to show that the doctrine of the unity of science and of the one universal language was incompatible with Tarski's theory of truth. Neurath thereupon suggested in the discussion which followed my lecture that Tarski's theories about the concept of truth must be untenable; and he inspired Årne Naess, who was also present, to undertake an empirical study of the usage of the word 'truth', in the hope of thus refuting Tarski.[26]

Carnap's reply to Popper's criticisms was contained in the book which Schilpp devoted to the study of his thought. Here he wrote:

The thesis of the unity of science, as Neurath and I maintained it, has nothing to do with the question of logical completeness. Rather, it was meant as a rejection of the division of empirical sciences into allegedly fundamentally separate fields, above all of the division into natural sciences and social sciences ('*Geisteswissenschaften*'), a division which was based on dualistic metaphysics prevailing at that time in Germany. In contrast to this dualistic conception, our thesis asserted that empirical science, with all its various fields, can be constructed on a uniform basis. Understood in this sense, I still maintain this thesis.[27]

Neither Popper's criticisms nor Carnap's reply can be accepted without reservation. Popper's critique seems justified with reference to Carnap's idea of the 'logical foundations of the unity of science', but hardly to Neurath's encyclopedism. Carnap's reply provides some valid, if incomplete, arguments in support of Neurath's position, but as a self-defence it remains unforthcoming.

There can be no doubt that Carnap's thesis of a 'unified science in a unified language' possessed the features attributed to it by Popper and that it contained the fallacies outlined by him. The same was true also of the positions held by Feigl,[28] Joergensen,[29] and for the most part by those members of the Circle more directly influenced by the common-sense background in philosophy favoured by Carnap's version of physicalism. Even in his already somewhat self-critical writings with respect to the theses he had maintained in *Erkenntnis* in the early Thirties – in, for example, 'Testability and Meaning'[30] and 'Logical Foundations of the Unity of Science'[31] – Carnap did not relinquish the notion that the language of physics was, in an important technical sense, the 'total language of science' which included the contents of all other languages. 'Physicalism', in Carnap's version, meant that every statement in the language of science was equivalent to some statement in the physical (or ('thing'-) language and could therefore be logically translated into this language without change of semantic content.[32] This was Carnap's well-known (and much disputed) doctrine of the 'reducibility', by means of definitions or other 'reduction statements', of all propositions in the language of science to protocol statements relating to observable properties (i. e. 'sense-data' or 'perception terms') of physical objects.[33]

In 'Logical Foundations of the Unity of Science', which was Carnap's contribution to the *Foundations* of the *Encyclopedia of Unified Science*, the thesis of the logical unity of the language of science and of the 'reducibility' of every meaningful language to the physical language was once again set out in detail. Here he maintained not only that "the class of observable thing-predicates is a sufficient reduction basis for the whole of the language of science"[34] (including biology, psychology, and the social sciences), but also that "the construction of one homogeneous system of laws for the whole of science is an aim for the future development of science".[35] The evidence – to his mind already gathered – for the possibility of unifying the language of science allowed us to "endeavour to develop science more and more in the direction of a unified system of laws".[36]

These extracts are sufficient demonstration of the difference in direction from the early Thirties on of the developing thought of Neurath and Carnap on the subject of 'unified science': Neurath's direction being pragmatic and sociological, Carnap's linguistic and logicistic.[37] Despite increasing difficulties, repeated exercises in self-criticism, and continual procrastinations, Carnap never ceased to work in the spirit of *Der logische Aufbau der Welt*. For all his life he was to continue pursuit of the unattainable goal of the construction of a complete, logically coherent, language of science, valid as a grammatical base for all the empirical disciplines, and, at the same time, strictly purged of all metaphysical elements.[38]

From the mid-Thirties on, Neurath's central concerns, often voiced as a warning accompanying the encyclopedic programme, were of a kind directly opposed to Carnap's logicistic aspirations. Not only did he view as dubious the possibility of a logical unification of the laws and methods of the various scientific practices but he also thought it necessary to regard the 'encyclopedic model' as opposite to the 'system model' (*Modell-System*) even at the level of the simple linguistic integration of scientific communication and its purification of metaphysics.[39] As he put it in *Foundations of the Social Sciences* with evident reference to Carnap's artificial languages

> A language scheme used for calculatory problems and a Universal Jargon are different in many respects. Not a few misunderstandings and difficulties seem to arise from mixing up both fields in our arguing.[40]

In Neurath's later writings the very distinction between science and metaphysics came to assume increasingly more problematic aspects, with a tendency to divide itself into subdistinctions and, at the end, to suffer a distinct watering-down. No attempt to purify the empirical vocabulary of metaphysical terms – as Neurath openly acknowledged in his 'Einheit der Wissenschaft als Aufgabe' of 1935 – could work on the basis of absolute and therefore radical criteria. The exercise of strict criticism might possibly lead at most to the elimination of the 'more egregious errors',[41] the 'coarser procedures', and the 'manifestly metaphysical' aspects.[42] But the metaphysical (or, conversely, scientific) nature of many terms would still remain undetermined, without it being possible to discontinue their use for this reason. In addition Neurath agreed that positive interaction had at times occurred in the past between science and metaphysics and also that a general theoretical hypothesis of the heuristic fruitfulness of some elements of metaphysics might well be maintained. At times too scientists had been inspired by far-fetched speculations, spurious conceits, and vague or arbitrary interpretations which had influenced the formation of important scientific theories. From the standpoint of encyclopedism, therefore, "the history of all these imaginations has to be regarded as a part of the history of the empiristic mosaic".[43]

Further still, it could not be denied that, alongside theology and truly 'speculative' metaphysics, there had been present in the development of Western thought – from Scholasticism to Descartes, Leibniz, and Kant – the element of 'rationalistic metaphysics'. This second type of metaphysics – aprioristic rationalism – had beyond doubt played a positive role of stimulus

(Stimulanz) in the development of the empirical sciences. Thus the relationship between rationalistic apriorism and empirical rationalism, between metaphysics and science, turned out to be an important aspect of sociology and psychology of science (*Forschungsbehavioristik*) and not merely an object of historiographical research or a part of the battleground of philosophical warfare.[44] Even from within religious thought and a theological milieu important contributions could arise towards logical and mathematical research or towards understanding of the methods of empirical enquiry – as proved to be the case not only with Poincaré, Duhem, and Le Roy, but also with Cantor, Brentano, Meinong, and Bolzano.[45]

3. *ESPRIT SYSTÉMATIQUE* VERSUS *ESPRIT DE SYSTÈME*: THE ENCYCLOPEDIC PARADIGM

In his contribution to the International Congress at Paris Neurath echoed worries expressed by Federigo Enriques about the danger of dogmatism and scholasticism present in logical empiricism[46] when he stated openly that the project of encyclopedism was opposed in the most direct way possible to the pseudorationalistic and logicistic 'illusion' that construction of '*the* system of science' was both possible and desirable:[47] "I have proposed", he wrote, "the term '*Enzyklopädie*' principally in opposition to the term '*System*', by which a kind of axiomatisation of the totality of science is meant".[48] In the same vein, in his essay published in Paris in 1936, 'L'Encyclopédie comme 'modèle'', Neurath maintained that:

> For an advocate of the empiricist attitude it is absurd to speak of a unique and total system of science. He must conceive his work as tending towards precision and systematisation *within an always variable framework which is that of an encyclopedia.* What we call 'encyclopedia', it seems to us, is nothing but a provisional assemblage of knowledge, not something still incomplete, but the totality of scientific matter now at our disposal. The future will produce new encyclopedias that will perhaps oppose ours; but for us it does not make any sense to speak of the 'complete encyclopedia' that could serve as a 'standard measure' for estimating the degree of perfection of the historically given encyclopedias. The encyclopedia that is the model of science is in no way unique and select; but we are dealing with encyclopedias each of which is a model of science, and one of which is applied at a definite period. The march of science progresses from encyclopedias to encyclopedias. It is this conception that we call *encyclopedism.*[49]

Here the epistemological category of 'encyclopedia' appears to be surpisingly close to that of 'thought-style' (*styl myślowy, Denkstil*'), used by Ludwik

Fleck, and even closer to that of 'paradigm' proposed by Thomas Kuhn at the height of the anti-positivistic 'revolution' in contemporary philosophy of science. 'Encyclopedism' was a theme Neurath developed fully in his disagreements with Popper. Popper had entered the debate at Paris with the argument that a non-systematic encyclopedia would have no appeal for specialists in the individual disciplines, and especially for physicists, since the system was for them the moving force behind research.[50] "*The* system", Neurath replied, "is the great scientific lie".[51] In works of slightly later date he argued, referring to d'Alembert and his critique of the *esprit de système*, for the necessity and the possibility of only a 'partial systematisation' *(partielle Systematisierung)* of interdisciplinary areas. It was important to take account of the historicity and the provisional character of developments and classifications within individual disciplines and not to start aprioristically, in Laplacean manner, from the metaphysical assumption of the logical unity and nomological consistency of the world of science.[52] As he wrote in *Le développement du Cercle de Vienne et l'avenir de l'empirisme logique*, published in Paris in 1935:

> We must unceasingly strive to draw attention to the fact that the Encyclopedia is bound to contain a great number of the imprecisions indirectly present in all the empirical sciences. We have no wish to be under any illusion about the imprecisions, lacunas, and contradictions by which every scientist is surrounded. It is not sufficient for us to have overcome the absolutism of systematic philosophy: we have to beware also of the pseudorationalism which treats the empirical sciences as if they were a matter of logical relations between clearly distinguished propositions, independent one of the other, and constructed on the basis of perfect definitions.[53]

The axiomatisation of the discursive procedures of science which was made possible by sophisticated techniques of modern symbolic logic could not, he said, extend beyond the bounds of a local rationality[54] and had in fact historically resulted only in partial or fragmentary formalisations.[55] All attempts to surpass these bounds met with insuperable problems connected with the ineradicable imprecision and indeterminacy of theoretical language and with the relativity and 'ambiguity' of all scientific propositions.[56] Any attempt at formalisation had to take account of the inadequate nature of experimental results which was bound always to be produced: these would always be inadequate with respect to the symbolic abstractions of theories.[57] It had to take account also of the ways in which the language itself employed by those conducting experiments was historically and sociologically conditioned: it was not in itself necessarily any more precise and univocal than everyday language.[58]

What on the contrary, Neurath argued, would be both possible and advantageous was the establishment of a network of 'cross-connections' *(Querverbindungen)*[59] between the partly systematised local areas, in such a way that scientific prediction might be able to put to use in any given case the 'instrumentation' of all relevant disciplines and theories.[60] These 'local systematisations' and cross-connections in the scientific disciplines did not necessarily exclude the possibility of contradictions existing between different theoretical areas: as the history of science had well demonstrated, theories which were incompatible with each other were perfectly capable of allowing valid predictions to be reached within their respective spheres.[61]

The *Encyclopedia*, he said, could be envisaged as a mosaic of varying patterns whose stones in some places overlapped one another and in others tended to come loose and required constant replacement. Alternatively it could be seen as an onion whose layers overlapped one another without being centered on any middle core.[62] In no case could its structure present a 'pyramidal' shape with hierarchical organisation of its internal levels.[63] 'Pyramidism', whose aim was to construct an image of science on harmonious and symmetrical lines, was no more than a relict of scholasticism. In contrast to the arrogance and pomposity of the 'architectonic' view of science, the new *Encyclopedia* aimed to present itself as an 'unpretentious'[64] programme refuting the speculative ideal of a 'natural order of the sciences'.[65] More than a 'retrospective synthesis' of scientific knowledge, the *Encyclopedia* was to be a means of the discovery of new paths and the identification of problems and unexplored possibilities of the integration of science.[66]

The *Encyclopedia* offered itself therefore not as an 'unhistorical ideal' but as a 'historical fact in the making'; it was not to be a mere catalogue or mausoleum but an instrument of the most 'vital activity' arising as a live intellectual force from the real needs of people. It would share then in the provisional nature and risks of any historical endeavour, would be influenced by fears and desires,[67] and might meet with the opposition of powerful social groups.[68] It was a 'provisional organisation of knowledge' which would be superseded in time by other encyclopedias, since it was in the nature of science to proceed from one encyclopedia to another, and not to develop as a logically consistent and unitary system.[69] Nor did it even form a kind of encyclopedic 'standard' valid for the time being, since an encyclopedia was no more than one model of science among the many between which a choice had to be made at any given point in history.[70]

It was the '*praxis* of life' (*Lebenspraxis*) – Neurath argued, adding an encyclopedic note to his earlier thoughts on conventionalism – rather than requirements of a logical kind which led scientists to adopt one encyclopedia in preference to another. By means of 'successive assimilations and eliminations' (*par une série d'assimilations et de rejects successifs*) the form of scientific knowledge emerged in each different age – we note again a surprising anticipation here of Thomas Kuhn's epistemological ideas – as certain 'uniform patterns of thought' (*des manières de penser uniformes*) from which the 'individual deviations' (*les écarts*) were few.[71] In the *praxis* of science

> it rarely happens that different competing conceptions involve one another in out-and-out confrontation arising from the discrepancy of their respective results.[72]

In this way it came about that, in one sense, scientists in each historical period found themselves obliged to choose between a plurality of encyclopedias logically opposed to one another and, in another, that their decisions were laid down and conditioned by practical factors and not by theoretical reasons, there being no general criterion to assist their choice from the standpoint of theory or even of the progression of science and society. The practical choices themselves of the scientists acted as 'touchstone' (*pierre de touche*) in the selection between individual encyclopedias in competition with one another.[73]

Each encyclopedia, Neurath argued in his last work on this subject, in as much as it was the necessarily provisional product of methodological decisions and practical choices made within the 'republic of scientists' (*Gelehrtenrepublik*), could not fail to be the result of a process of 'orchestration', that is, of a series of 'successive adaptations and compromises'[74] leading to general agreement among scientists. The *Encyclopedia* was therefore the expression of a spirit of intellectual tolerance and democratic cooperation opposed to any form of 'imperialism' or doctrinal totalitarianism. For this reason Neurath not only proclaimed his relinquishment of the very term 'positivism' (and logical positivism), which he now considered to be compromised by anti-pluralistic positions, but argued also that it was necessary to 'cut all links' with nineteenth century positivism, especially with the dogmatism and authoritarian moralism of the 'Comtean church'.[75] The basic aim of the *Encyclopedia* was to provide a 'platform' for democratic cooperation between citizens of the 'republic of scientists', between scientists and the 'man in the street', and between nation and nation.[76] In the spirit of American pragmatism – to borrow Dewey's words in his contribution, 'Unity of Science as a Social Problem', in volume I of *Foundations of the Unity of Science*[77] – the encyclopedic idea based itself on the educative efficacy of a 'comprehensive scientific attitude' opposed to dogmatism, prejudice, absolutism, and the use of force to protect special (including nationalistic and racial) interests.[78]

For these reasons too encyclopedism was opposed to traditional European metaphysics, most especially to the German idealism which was responsible for the diffusion through Europe of an atmosphere of political irrationality, intolerance, and totalitarism.[79] "Metaphysics divide", Neurath declared, "science unites". While the scientists who participated in the process of the 'encyclopedic unification of science' were devoting themselves to the creation of a kind of 'scientific republic of work' (*Gelehrtenrepublik der Arbeit*), the metaphysical philosophers were acting like the lords of San Gimignano "locked in their towers at dead of night, seeking to protect themselves against each other by raising their towers ever higher; but, because it is dark, they feel afraid and sing to themselves. (Freud too depicted his wandering philosophers as singing in the dark of the wood). It may be that this serves to alleviate the fear, but it is not thus that the new day dawns in the world".[80]

4. THE EPISTEMOLOGICAL PRIORITY OF SOCIOLOGY: A CRITICISM OF THE 'COVERING-LAWS-MODEL' OF EXPLANATION

Portrayal of the encyclopedic programme as close to an open and tolerant 'scientific empiricism' (such as would include, as well as 'logical empiricism', all the trends in relativism, such as pragmatism, instrumentalism, and conventionalism)[81] ended, however, by having effect also on the 'physicalist' premises of Neurath's thought. Although an indisputable philosophical primacy of the categories and language of physics had been entailed in the original formulations of physicalism, this primacy was greatly diminished in Neurath's later writings to the extent that its position became altogether precarious, if not actually abandoned.

Previously, in 1934, at the Precongress at Prague, Neurath had observed that the physicalist unification of the language of science did not involve the thesis of the deducibility of the laws of philosophy, psychology, and sociology from the laws of physics.[82] He reinforced and amplified this warning in his discussion with Å. Petzäll on 'Physikalismus und Erkenntnisforschung':

the physicalist thesis that it was possible to reduce (*zurückführen*) the terms of all the empirical disciplines to the terms of physics was not equivalent to the thesis that "all statements and laws of the individual scientific disciplines can be reduced to the statements and laws of physics".[83] Notwithstanding Carnap's endeavours, the very reducibility of all the terms of one discipline to the terms of another – for example, those of sociology to those of physics – was an open question requiring further investigation.[84]

He warned too of the great danger present in considering the unitary language of physicalism as the lexical base of all the sciences:

> The danger is that the logical clarity of formulas in physics may be taken as a paradigm for all the sciences.[85]

It was necessary therefore to take systematic notice of the imprecisions and gaps in theory with which the practical activity of the scientist was beset:

> Quite properly we shall be forced to take more systematic account of these 'imprecisions', whose full revelation is just as important as the renunciation of 'atomic propositions' or of other supposed bases for theoretical constructions of a metaphysical nature. *Imprecision* and *multiplicity* are intrinsic to the empirical sciences.[86]

Social scientists, he remarked,

> sometimes think of physics and astronomy as of an El Dorado of exactness and definiteness, and they assume, frequently, that in this field any kind of contradictions are fatal to hypotheses.[87]

They thought there must be an essential difference in the certainty of the foundations and results of the different disciplines, but did not realise that imprecision, insufficient information, and contradictions were present even within the natural sciences and that, in some cases, these were precisely the areas where prediction was less readily possible.

It might, for instance, prove more difficult to predict the weather than the result of elections taking place next month or, again, it might be possible to forecast the behaviour of living beings with greater precision than the movements of heavenly bodies such as meteors.[88] The prediction of earthquakes was not in itself more reliable than the prediction of wars or revolutions.[89] In spite of this the opinion had still prevailed in logical empiricism that analysis of the procedures of physics could lead also to definition of the methodological structure of other disciplines. But this could not hold as a general rule and was the reason why

> the project for the unification of science does not start from the idea that physics can be considered *tout court* as a model for all the other sciences'.[90]

Laboratory experiments, for example, a fundamental procedure in physics, were of much more limited importance in biology, to say nothing of psychology and sociology.[91] The very possibility of applying mathematical means in a discipline such as sociology was subject to closely defined limits: sociological generalisations held normally only 'for relatively complex structures' (*für relativ komplexe Gebilde*) and only within a limited scope of space and time.[92] Given, for example, that certain constant quantitative relations had been established between variations in the relative incidence of illiteracy and crime in certain urban centres within a defined area and over a defined period of time, then it would be possible to conjecture that, according to a certain scale, these quantitative relations would hold also for other cities within the same area, but one would be extremely wary of supposing that they could hold for different times or for different countries or for cities which might conceivably be discovered on Mars.[93]

In the light of this it might well appear that a basic procedural difference existed between physics and sociology and that no more than a vague approximation to the model of causal explanation employed in physics could be achieved in other disciplines. Here, however, Neurath turned once more to classical conventionalism (Mach especially) and, reversing the normal tenor of the neo-empiricist othodoxy and its falsificationist variant, established precise reasons for his opposition to classical positivistic causalism in a clearly defined critique of the deductive-nomological model of scientific explanation[94] which has claim to stand as one of the high points of his philosophical thought.

No longer now did he appeal to a principle of philosophical tolerance to legitimise, beside the deductive-nomological model, statistical generalisation or the use of 'existential propositions'. Even the so-called 'laws' of physics, he declared, could be considered valid only within limited areas and defined periods of time. They were generalisations formally and logically no different from the statistical generalisations which were typically the means towards sociological predictions. That the so-called 'laws of physics' were normally formulated as universal laws of nature was due solely to the technical artifice of omitting clear indication of the spatial and temporal limitations to the validity of their forecasts or of introducing limiting clauses, such as *ceteris paribus* or *rebus sic stantibus*, of a vague or indeterminate nature.[95]

Encyclopedism, Neurath said, did not start from the assumption of the validity of 'natural laws'.[96] On the contrary its point of departure was the recognition that the 'weak generalisations' (*weniger anspruchsvolle Verallgemeinerungen*) regularly employed in the social sciences were employed also in physics, the difference being merely one of degree (*nur gradweise Unterschied*) and not one of principle. In physics no less than in sociology experiments based on a strictly limited number of examples were extended to an equally strictly limited number of further cases.[97] In sociology a case in point would be the number of nations, countries, social systems, inhabitants of the planet etc. But, as Neurath made especially clear, something of the same kind occurred also in astronomy, where Kepler's laws, for example, had been put foward on the basis of particular observations valid only to the extent of a limited number of planets.[98]

All scientific predictions, those of physics included,

> are based on the fact that certain quantitative relations remain sufficiently constant notwithstanding undeniable changes in general conditions. And while individual experiments are extrapolated in sociology only to a limited degree for the reason that the assumption is made that the normal variations (*übliche Änderungen*) of general conditions change the quantitative relations between phenomena, in physics or in chemistry the supposition is made that a large number of the rules used to formulate predictions can be applied without the necessity of taking into account any practically important quantitative changes.[99]

It followed that what could be maintained was that

> the rules of prediction are in individual cases capable of different extension and it is for this reason that predictions relating to various phenomena or objects, such as cities, forests, uranium deposits, water reserves, can be treated in analogous manner. In the light of this understanding the so-called 'laws' of physics cover only a scope of limited, even if wide, validity, and the bounds of the validity of predictions however much they are taken as underlying assumptions, are normally left imprecise.[100]

Such relativisation (*Relativisierung*) and historicising (*Historisierung*) of physics, he added, was directly prompted by the teaching of Mach, who had been first to wake up to the danger of the exaggerations (*Überschwenglichkeiten*) of the predictive capacity of physics, and had questioned, for example, what validity the principle of inertia would retain, if all the stars were one day to move simultaneously out of order.[101]

Once the advantage of this theoretical approach was recognised, then,

Neurath argued, it followed that the possibility of using quantitative hypotheses of a physical, biological, or sociological kind for predictive purposes made it necessary that individual predictions should be attached to a general prediction of a 'cosmological' nature.[102] Hypotheses should, that is, relate to the whole context of 'general conditions' of a spatial and temporal kind (*kosmologische Gesamtlage*) from which the individual predictions drew their relative validity and the individual predictions should at all times be accompanied by further *ad hoc* assumptions about the constancy of the 'cosmic' context of reference.[103]

The deductive-nomological model of scientific explanation and prediction which was erroneously held to operate in physics could not therefore be taken, as Popper claimed (and as Hempel was to argue in a debate important for the development of logical empiricism), to stand as the general model of scientific explanation and to be for that reason the yardstick of scientific quality in all the empirical disciplines, including sociology. If anything, the 'weak generalisations' of sociology formed the only model which could legitimately be taken as a general basis of empirical prediction. But it is evident that this basis did not constitute for Neurath a 'deductive model' enabling transition to take place from definition of 'statistical laws' to logical inference of the probability of empirical events (as Hempel was to argue in the Seventies).[104] For Neurath prediction was always *inference of the particular from the particular by means of statistical and inductive extrapolation*. And at the root of inductive statistical extrapolation there operated, along with techniques of a logical nature, techniques of decision which could not be formalised, as Carnap would claim, by any 'inductive logic'. An element of judgment was inevitably involved not only in the criteria – or classes of reference – of statistical extrapolation[105] but also in the clauses such as *ceteris paribus* and *rebus sic stantibus* which conventionally supposed the constancy or irrelevance of the variations in general conditions which in fact formed the 'cosmological' context of prediction.[106]

The 'major premise' of scientific explanation could never – except through an arbitrary assumption – take the form of a universal statement: neither the strictly universal one of the 'laws of nature' claimed by Popper nor that of a statistical 'law' from which it was nevertheless possible logically to 'deduce' the probability of an event. Here we find, therefore, the entire covering-laws-model of explanation and prediction receiving from Neurath clear and incisive anticipatory refutation.

In the light of these premises it is hardly surprising that Neurath should find the exemplary model of methodological correctness, to which all other disciplines should adapt themselves, in the *praxis* not of physical but of sociological research:

> while discerning sociologists normally attempt to indicate the limits of validity in their generalisations, this is not the practice of physicists, who rarely disclose the fact that all their laws should be formulated in harness with certain limitative clauses of a 'cosmological' kind.[107]

It was physicists far more than sociologists who showed themselves unaware of the relative and historical nature of all forms of knowledge and empirical prediction, and consequently of the impracticability of 'absolutistic' criteria of verification or falsification of hypotheses.[108] Physicists,

> often finding themselves in agreement on certain fundamental issues, rarely recognise that in the final analysis it is the entire ensemble of their hypotheses which is in dispute and not merely the limited area with which they are more directly occupied.[109]

Physicists were more ready than social scientists to overlook the fact that

> there are times in the history of science when it is necessary to take note that the totality of hypotheses may always be placed in doubt.[110]

It was precisely this recognition of the limits present in physical as well as sociological research – and recognition above all that these limits were identical – which was to be taken as a basis for the assertion of unity between the natural and the social sciences. In a reversal of the classical positions of reductionism and dogmatic empiricism, Neurath maintained that a 'discerning' sociologist, when pleading the cause of the unity of science, would have to

> be concerned much less with showing that sociology is able to obtain the same excellent results as those achieved by other sciences than with giving prominence to the fact that the bounds already recognised to exist for sociology exist also for all the other sciences.[111]

The novelty of this position and its startling originality in the context of Viennese empiricism found its confirmation and greatest advertisement in the charge of 'historicism' which Popper levelled against it. In *The Poverty of Historicism*, at the end of a chapter devoted to the critique of the anti-naturalistic theses of historicism, Popper had no hesitation in including Neurath alongside Mannheim among the 'historicists'. In his desire to bring to light the unity of physics and sociology the 'historicist' Neurath advances the proposal:

that instead of vainly attempting to follow in sociology the example of physics, and to search for universal sociological laws, it would be better to follow in physics the example of a historicist sociology, i.e. to operate with laws which are limited to historical periods.[112]

Against this 'historicist' position Popper maintained that, if physicists were to add to the formulation of 'natural laws' clauses which would restrict their validity to the 'present cosmological period', then they would give evidence, not of their commendable scientific prudence, but of their failure to understand how science actually functions.[113] For

it is an important postulate of scientific method that we should search for laws with an unlimited realm of validity. If we were to admit laws that are themselves subject to change, change could never be explained by laws. [...] And it would be the end of scientific progress; for if unexpected observations were made, there would be no need to revise our theories: the *ad hoc* hypothesis that the laws have changed would 'explain' everything. These arguments hold for the social sciences no less than for the natural sciences.[114]

Beyond the arguments employed by Popper against historicism in sociology, it is worth noting this singular instance of the 'positivistic' sociologist Neurath here being pilloried for 'historicism' and 'relativism' – and thus of antipositivism – by the very philosopher of science who is generally credited with having first denounced the positivist dogmas of the Vienna Circle.[115]

Despite Popper's harsh (and in some measure contradictory) arguments, it remains to say in conclusion that Neurath's encyclopedic project was hardly a 'monument' to methodological foolishness and metaphysical dogmatism. Rather, the *neue Enzyklopädie* was, in its basic theoretical premises and in its considered formulations, testimony to a clear perception of the possibilities and limitations of a modern science understood as a self-reflexive historical undertaking, as the product of collaboration and free exchange of ideas, and as pragmatically directed towards the regeneration of culture and society. It was a 'monument' to an ardent faith in scientific rationality and, at the same time, to intellectual austerity, modesty, tolerance, and, not least, epistemological scepticism. As Neurath wrote in one of his last works

such encyclopedism is the expression of a certain skepticism which objects not only to metaphysical speculations but also to overstatements within the field of empirical sentences.[116]

NOTES

[1] On the series of meetings and Congresses organised by the Vienna Circle and particularly by Neurath see: J. Joergensen, *The Development of Logical Empiricism*, Chicago: The University of Chicago Press, 1951, pp. 43–8; F. Barone, *Il neopositivismo logico*, pp. 9–13; F. Barone, 'Introduzione' to O. Neurath, *Il Circolo di Vienna e l'avvenire dell'empirismo logico*, pp. 9–13. As to the Paris Congress, see particularly O. Neurath, 'Erster Internationaler Kongress für Einheit der Wissenschaft in Paris 1935', [1935c], pp. 377–406; see in addition: *Actes du Congrès international de philosophie scientifique, Sorbonne, Paris 1935*, in *Actualités Scientifique et Industrielles*, Paris 1936.

[2] Cf. F. Barone, 'Introduzione' to O. Neurath, *Il Circolo di Vienna e l'avvenire dell'empirismo logico*, p. 9.

[3] For information concerning this Institute see: the Editors' Note to the 'Postscript' to J. Joergensen, *The Development of Logical Empiricism*, p. 100; O. Neurath, 'Die neue Enzyklopädie des wissenschaftlichen Empirismus', *Scientia*, *62* (1937), now in *GpmS*, pp. 804–9, English trans. *PP*, pp. 192–6; G. Statera, *Logica, linguaggio e sociologia. Studio su Otto Neurath e il neopositivismo*, pp. 29–30, 127.

[4] On the editorial aspects of the *International Encyclopedia of Unified Science* see: F. Barone, *Il neopositivismo logico*, pp. 352–4; C. Morris, 'On the History of the International Encyclopedia of Unified Science', *Synthese*, *12* (1960), pp. 517–21; O. Neurath, et al., *Zur Enzyklopädie der Einheitswissenschaft*, no. 6, Den Haag: Van Stockum & Zoon, 1938, English trans. O. Neurath (ed.), *Unified Science*, ed. by B.F. McGuiness and H. Kaal, Dordrecht and Boston: D. Reidel, 1987.

[5] Cf.: O. Neurath, 'Die wissenschaftliche Sprache', in *Unity of Science Forum*, supplement to *Synthese*, 1938, now in *GpmS*, pp. 846–7; O. Neurath, 'Die neue Enzyklopädie', in O. Neurath, et al., *Zur Enzyklopädie der Einheitswissenschaft*, no. 6, Den Haag: Van Stockum & Zoon, 1938, now in *GpmS*, p. 869, English trans. in O. Neurath (ed.), *Unified Science*.

[6] Among the many references in Neurath's writings to G. Itelson see for instance *GpmS*, pp. 784, 803, 861, 885, 905, 998.

[7] Cf. O. Neurath, 'Die neue Enzyklopädie des wissenschaftlichen Empirismus', [1937c], *GpmS*, p. 801, *PP*, pp. 189–90.

[8] Cf.: O. Neurath, 'L'encyclopédie comme 'modèle'', [1936d], p. 201, *PP*, p. 158; O. Neurath, 'Physikalismus und Erkenntnisforschung', II, [1936f], *GpmS*, p. 757, *PP*, p. 168; O. Neurath, 'Unified Science as Encyclopedic Integration', [1938d], p. 5. See also G. Statera, *Logica, linguaggio e sociologia. Studio su Otto Neurath e il neopositivismo*, pp. 119–22.

[9] Cf.: O. Neurath, 'Une Encyclopédie internationale de la science unitaire', [1936c], in *Actes du Congrès international de philosophie scientifique, Sorbonne, Paris 1935*, II, pp. 58–9, English trans. *PP*, p. 143; O. Neurath, 'L'encyclopédie comme 'modèle'', [1936d], p. 200, *PP*, p. 157; O. Neurath, 'Unified Science as Encyclopedic Integration', [1938d], pp. 15–9; O. Neurath, 'Unified Science and Its Encyclopedia',

[1937a], pp. 274 ff, *PP*, pp. 179 ff; O. Neurath, 'Einheitswissenschaft als empiristische Synthese', [1938], *GpmS*, p. 828; O. Neurath, 'Die neue Enzyklopedie', [1938c], *GpmS*, pp. 864–5.
[10] Cf. O. Neurath, 'Une Encyclopédie internationale de la science unitaire', [1936c], p. 56, *PP*, p. 140.
[11] O. Neurath, 'Einheit der Wissenschaft als Aufgabe', [1935], p. 16, *PP*, p. 115.
[12] O. Neurath, 'Une Encyclopédie internationale de la science unitaire', [1936c], pp. 55–6, *PP*, p. 140.
[13] O. Neurath, 'Radikaler Physikalismus und 'Wirkliche Welt'', [1934], pp. 353–4, *PP*, p. 106; O. Neurath, 'Einheit der Wissenschaft als Aufgabe', [1935], pp. 16–7, *PP*, p. 115; O. Neurath, *Le développement du Cercle de Vienne et l'avenir de l'empirisme logique*, [1936], pp. 53–7; O. Neurath, 'L'encyclopédie comme 'modèle'', [1936d], pp. 199–201, *PP*, pp. 156–8.
[14] O. Neurath, 'Universal Jargon and Terminology', [1941], pp. 127–48, passim, *PP*, pp. 213–229, passim.
[15] O. Neurath, 'L'encyclopédie comme 'modèle'', [1936d], pp. 198–9, *PP*, pp. 155–6; O. Neurath, 'Unified Science and Its Encyclopedia', [1937a], pp. 270–1, *PP*, pp. 175–6.
[16] O. Neurath, 'Universal Jargon and Terminology', [1941], pp. 128–9, 131, *PP*, pp. 214–5, 216.
[17] O. Neurath, 'Physikalismus', [1931d], *GpmS*, p. 418, *PP*, p. 53.
[18] A thesis often found in Neurath's theoretical writings of 1931, such as 'Physikalismus', *Empirische Soziologie*, 'Soziologie im Physikalismus'.
[19] O. Neurath, *Einheitswissenschaft und Psychologie*, [1933], *GpmS*, pp. 587–610.
[20] The charge of 'implicit realism' was levelled by G. Statera, *Logica, linguaggio e sociologia. Studio su Otto Neurath e il neopositivismo*, pp. 49, 82, 102; and by F. Barone, *Il neopositivismo logico*, pp. 333–4.
[21] Cf.: R. Carnap, 'Die physikalische Sprache als Universalsprache der Wissenschaft', *Erkenntnis, 2* (1931), pp. 432–65, English trans. London: Kegan Paul, 1932, under the title *The Unity of Science*; R. Carnap, *Philosophy and Logical Syntax*, London: Kegan Paul, Trench, Trubner & Co., 1935; R. Carnap, 'Über die Einheitssprache der Wissenschaft: Logische Bemerkungen zum Projekt einer Enzyklopädie', in *Actes du Congrès international de philosophie scientifique, Sorbonne, Paris 1935*, II, pp. 60–70; R. Carnap, 'Einheit der Wissenschaft durch Einheit der Sprache', in *Travaux du IXe Congrès international de philosophie, Congrès Descartes*, IV, in *Actualités scientifiques et industrielles*, no. 533, Paris 1937; R. Carnap, 'Logical Foundations of the Unity of Science', in *International Encyclopedia of Unified Science*, pp. 42–62.
[22] Cf. O. Neurath, 'The Orchestration of the Sciences by the Encyclopedism of Logical Empiricism', *Philosophy and Phenomenological Research, 6* (1945–46), pp. 496–508, now also in *PP*, pp. 230–42.
[23] In P.A. Schilpp (ed.), *The Philosophy of Rudolf Carnap*, pp. 183–226.
[24] Tarski's essay was originally published in Polish in 1933, followed by the German translation in 1935 (in *Studia Philosophica*).
[25] K.R. Popper, 'The Demarcation between Science and Metaphysics', p. 201.
[26] Ibid, loc. cit., note 24. Cf.: Å. Naess, *'Truth' as conceived by those who are not Professional Philosophers*, Oslo: I Kommisjon hos J. Dubwad, 1938; Å. Naess, *Four*

Modern Philosophers. Carnap, Wittgenstein, Heidegger, Sartre, Chicago: The University of Chicago Press, 1965, pp. 11–66. See in addition Popper's references to his own disagreement about the *International Encyclopedia* in 'Memories of Otto Neurath', *ES*, p. 56.

[27] R. Carnap, 'K. Popper on the Demarcation between Science and Metaphysics', in P.A. Schilpp (ed.), *The Philosophy of Rudolf Carnap*, p. 880.

[28] Cf.: H. Feigl, 'Unity of Science and Unitary Science', in H. Feigl and M. Brodbeck (eds.), *Readings in the Philosophy of Science*, New York: Appleton-Century-Crofts, Inc., 1963, pp. 382–4; H. Feigl, 'Physicalism, Unity of Science and the Foundations of Psychology', in P.A. Schilpp (ed.), *The Philosophy of Rudolf Carnap*, pp. 227–67; H. Feigl, 'The Origin and Spirit of Logical Empiricism', in P. Achinstein and S.F. Barker (eds.), *The Legacy of Logical Positivism*, pp. 3–24.

[29] Cf.: J. Joergensen, *The Development of Logical Empiricism*, pp. 76–83; J. Joergensen, 'Empiricism and the Unity of Science', *Erkenntnis*, *8* (1939–40), pp. 411–23.

[30] Cf. R. Carnap, 'Testability and Meaning', passim.

[31] Cf. R. Carnap, 'Logical Foundations of the Unity of Science', pp. 42–62.

[32] Cf. R. Carnap, *Philosophy and Logical Syntax*, p. 89. See also F. Barone, *Il neopositivismo logico*, pp. 31–2.

[33] Cf. R. Carnap, 'Testability and Meaning', pp. 67, 69–72. For a brilliant critique of this doctrine see K.R. Popper, 'The Demarcation between Science and Metaphysics', pp. 195–212. See also: C.G. Hempel, 'Implications of Carnap's Work for the Philosophy of Science', in P.A. Schilpp (ed.), *The Philosophy of Rudolf Carnap*, pp. 685–709; F. Barone, *Il neopositivismo logico*, pp. 416–33.

[34] R. Carnap, 'Logical Foundations of the Unity of Science', p. 60.

[35] Ibid., p. 61.

[36] Ibid., loc. cit.

[37] Cf. C.G. Hempel, 'Logical Positivism and the Social Sciences', in P. Achinstein and S.F. Barker (eds.), *The Legacy of Logical Positivism*, pp. 174–5 ("[Neurath] cautioned against overestimating the ideal of linguistic precision and held that artificially constructed formal languages, on which Carnap relied extensively in his work, had only limited value for the clarification of philosophical issues. Accordingly, he did not conceive of the unitary physicalistic language of science as having a precise vocabulary and a rigorously specifiable formal structure").

[38] See Carnap's tentative replies to the criticisms of C.G. Hempel, E. Nagel and W.V.O. Quine in P.A. Schilpp (ed.), *The Philosophy of Rudolf Carnap*, pp. 685–709 and 958–66, 785–825 and 989–95, 385–406 and 915–22. Concerning the different stages of Carnap's philosophical development see the accurate and critical reconstruction by C.G. Hempel, in 'Logical Positivism and the Social Sciences', pp. 174–85.

[39] O. Neurath, 'Pseudo', [1935a], pp. 354–5, *PP*, p. 122.

[40] O. Neurath, *Foundations*, [1944a], p. 3.

[41] O. Neurath, 'Einheit der Wissenschaft als Aufgabe', [1935], p. 21, *PP*, p. 118.

[42] O. Neurath, *Le développement du Cercle de Vienne et l'avenir de l'empirisme logique*, [1936], p. 55.

[43] O. Neurath, 'Unified Science as Encyclopedic Integration', [1938d], pp. 12–3, 21.

[44] O. Neurath, 'Die neue Enzyklopädie des wissenschaftlichen Empirismus', [1937c], *GpmS*, p. 810, *PP*, p. 197.

[45] O. Neurath, 'Wege der wissenschaftlichen Weltauffassung', [1930/31], p. 120, *PP*, p. 43; O. Neurath, *Le développement du Cercle de Vienne et l'avenir de l'empirisme logique*, [1936], pp. 33–41; O. Neurath, 'Der Logische Empirismus und der Wiener Kreis', [1936e], *GpmS*, p. 742. On the so called 'Neurath-Haller thesis', according to which there is a direct and strong link, via Brentano, between Catholic Scholasticism and the Vienna Circle's 'typical Austrian philosophy', see: R. Haller, 'Zur Historiographie der Österreichischen Philosophie', in J.C. Nyìri (ed.), *Austrian Philosophy. Studies and Texts*, München: Philosophia, 1986, pp. 41–53; J.C. Nyíri, 'The Austrian Element in the Philosophy of Science', ibid., pp. 141–6; B. Smith, 'Austrian Origins of Logical Positivism', in B. Gower (ed.), *Logical Positivism in Perspective*, London: Croom Helm, 1987, pp. 38–9, 42–8. See also the Carnap-Neurath correspondence, and especially Neurath's letter to Carnap of 15th January 1943, in the 'Carnap Collection', no. 102–55-02, Hillman Library, University of Pittsburgh (Pa).

[46] Cf. O. Neurath, 'Erster Internationaler Kongress für Einheit der Wissenschaft in Paris 1935', [1935c], p. 386. Philipp Frank, too, shared this position; cf.: O. Neurath, 'L'encyclopedie comme 'modèle'', [1936d], p. 191, *PP*, pp. 148–9; O. Neurath, 'Erster Internationaler Kongress für Einheit der Wissenschaft in Paris 1935', [1935c], p. 393.

[47] O. Neurath, 'Einzelwissenschaften, Einheitswissenschaft, Pseudorationalismus', [1936a], *GpmS*, p. 708, *PP*, p. 137; O. Neurath, 'Le développement du Cercle de Vienne et l'avenir de l'empirisme logique', [1936], p. 54; O. Neurath, 'L'encyclopédie comme 'modèle'', [1936d], pp. 187, 190, *PP*, pp. 145, 148.

[48] O. Neurath, 'Physikalismus und Erkenntnisforschung', II, [1936f], *GpmS*, p. 757, *PP*, p. 168.

[49] O. Neurath, 'L'encyclopédie comme 'modèle'', [1936d], p. 188, *PP*, p. 146.

[50] O. Neurath, 'Erster Internationaler Kongress für Einheit der Wissenschaft in Paris', [1935c], pp. 389–90.

[51] O. Neurath, 'Einheit der Wissenschaft als Aufgabe', [1935], p. 17 ("'*Das' System ist die große wissenschaftliche Lüge*"), *PP*, p. 116; O. Neurath, 'The Departmentalization of Unified Science', *Erkenntnis*, 7 (1937–38), pp. 244–6, English trans. *PP*, pp. 203–4.

[52] O. Neurath, 'L'encyclopédie comme 'modèle'', [1936d], pp. 188, 201, *PP*, pp. 145, 158; O. Neurath, 'Physikalismus und Erkenntnisforschung', II, [1936f], *GpmS*, p. 757, *PP*, p. 168; O. Neurath, 'Einheit der Wissenschaft als Aufgabe', [1935], p. 17, *PP*, p. 116; O. Neurath, 'Einzelwissenschaften, Einheitswissenschaft, Pseudorationalismus', [1936a], *GpmS*, p. 704, *PP*, p. 133.

[53] O. Neurath, *Le développement du Cercle de Vienne et l'avenir de l'empirisme logique*, [1936], p. 54; O. Neurath, 'Universal Jargon and Terminology', [1941], pp. 128–31 ("Sometimes a symbolism may even conceal the ambiguity of certain explanations and lull to sleep the attention of scientists who are accustomed to rely on symbolic argument", ibid., p. 127), *PP*, p. 213.

[54] O. Neurath, *Le développement du Cercle de Vienne et l'avenir de l'empirisme logique*, [1936], pp. 54, 56.

[55] O. Neurath, 'Une Encyclopédie internationale de la science unitaire', [1936c], p. 55, *PP*, pp. 139–40.

[56] O. Neurath, 'Unified Science and Its Encyclopedia', [1937a], pp. 268, 276, *PP*, pp. 174, 180; O. Neurath, 'Universal Jargon and Terminology', [1941], p. 130, *PP*, p.

174, 180; O. Neurath, 'Universal Jargon and Terminology', [1941], p. 130, *PP*, p. 216.

[57] O. Neurath, 'L'encyclopédie comme 'modèle'', [1936d], p. 190, *PP*, p. 148.

[58] O. Neurath, 'Einzelwissenschaften, Einheitswissenschaft, Pseudorationalismus', [1936a], *GpmS*, pp. 704–5, 707–8, *PP*, pp. 134–5, 136–7; O. Neurath, 'Une Encyclopédie internationale de la science unitaire', [1936c], p. 55, *PP*, pp. 139–40; O. Neurath, 'L'encyclopédie comme 'modèle'', [1936d], p. 189, *PP*, p. 147.

[59] O. Neurath, 'Einzelwissenschaften, Einheitswissenschaft, Pseudorationalismus', [1936a], *GpmS*, p. 705 , *PP*, p. 133; O. Neurath, *Le développement du Cercle de Vienne et l'avenir de l'empirisme logique*, [1936], p. 56; O. Neurath, 'Der Logische Empirismus und der Wiener Kreis', [1936e], *GpmS*, p. 746; O. Neurath, 'L'encyclopédie comme 'modèle'', [1936d], p. 197, *PP*, p. 155; O. Neurath, 'Die neue Enzyklopädie des wissenschaftlichen Empirismus', [1937c], *GpmS*, p. 802, *PP*, p. 191.

[60] O. Neurath, 'Une Encyclopédie internationale de la science unitaire', [1936c], p. 56, *PP*, p. 141.

[61] O. Neurath, 'L'encyclopédie comme 'modèle'', [1936d], p. 199, *PP*, p. 156; O. Neurath, 'The Orchestration of the Sciences by the Encyclopedism of Logical Empiricism', [1945/46], p. 498, *PP*, p. 156.

[62] O. Neurath, 'Unified Science as Encyclopedic Integration', [1938d], pp. 24–5; O. Neurath, 'The Departmentalization of Unified Science', [1937/38], p. 246, *PP*, p. 204; O. Neurath, 'Die neue Enzyklopädie des wissenschaftlichen Empirismus', [1937c], *GpmS*, p. 807, *PP*, p. 194.

[63] O. Neurath, 'The Departmentalization of Unified Science', [1937/38], pp. 244–6, *PP*, pp. 203–4; O. Neurath, 'The Social Sciences and Unified Science', *Erkenntnis*, *8* (1939–40), Appendix, pp. 430–2, now also in *PP*, pp. 209–12; O. Neurath, 'Wissen und Sein', German trans. from Dutch (*Algemeen Nederlands Tijdschrift voor Wijsbegeerte en Psychologie*, *31*, 1937/38, no. 5), *GpmS*, p. 836.

[64] O. Neurath, 'The Departmentalization of Unified Science', [1937/38], p. 246, *PP*, p. 204; O. Neurath, 'Der Logische Empirismus und der Wiener Kreis', [1936e], *GpmS*, p. 746.

[65] O. Neurath, 'The Departmentalization of Unified Science', [1937/38], p. 245, *PP*, p. 203.

[66] O. Neurath, 'Une Encyclopédie internationale de la science unitaire', [1936c], p. 55, *PP*, p. 140.

[67] O. Neurath, 'L'encyclopédie comme 'modèle'', [1936d], pp. 188, 200, *PP*, pp. 146, 157–8; O. Neurath, 'Unified Science and Its Encyclopedia', [1937a], p. 277, *PP*, p. 181; O. Neurath, 'Die neue Enzyklopädie des wissenschaftlichen Empirismus', [1937c], *GpmS*, p. 808; O. Neurath, 'Unified Science as Encyclopedic Integration', [1938d], pp. 25–6.

[68] O. Neurath, *Le développement du Cercle de Vienne et l'avenir de l'empirisme logique*, [1936], p. 57.

[69] O. Neurath, 'L'encyclopédie comme 'modèle'', [1936d], p. 188, *PP*, p. 146.

[70] Ibid., loc. cit.

[71] Ibid., p. 199, *PP*, p. 157.

[72] Ibid., loc. cit.

[73] Ibid., p. 200, *PP*, p. 157.

[74] O. Neurath, 'Universal Jargon and Terminology', [1941], p. 129, *PP*, p. 214; O. Neurath, 'The Orchestration of the Sciences by the Encyclopedism of Logical Empiricism', [1945/46], p. 508, *PP*, p. 242.
[75] Ibid., p. 501, *PP*, p. 235.
[76] Ibid., p. 508, *PP*, p. 242; O. Neurath, 'Die neue Enzyklopädie des wissenschaftlichen Empirismus', [1937c], *GpmS*, p. 810, *PP*, pp. 197–8; O. Neurath, 'Die neue Enzyklopädie', [1938c], *GpmS*, pp. 867–8.
[77] Cf. J. Dewey, 'Unity of Science as a Social Problem', in *International Encyclopedia of Unified Science*, vol. I, no. 1, pp. 34, 37–8.
[78] O. Neurath, 'The Orchestration of the Sciences by the Encyclopedism of Logical Empiricism', [1945/46], pp. 504–5, 508, *PP*, pp. 238, 242; O. Neurath, 'The Departmentalization of Unified Science', [1937/38], p. 246, *PP*, p. 204.
[79] O. Neurath, 'The Orchestration of the Sciences by the Encyclopedism of Logical Empiricism', [1945/46], p. 504, *PP*, p. 238.
[80] O. Neurath, *Einheitswissenschaft und Psychologie*, [1933], *GpmS*, p. 610.
[81] O. Neurath, 'Die neue Enzyklopädie des wissenschaftlichen Empirismus', [1937c], *GpmS*, p. 801, *PP*, pp. 189–90.
[82] O. Neurath, 'Einheitswissenschaft', in *Actes du Huitième Congrès international de philosophie à Prague, 1934*, Prag: Orbis, 1936, now also in *GpmS*, pp. 762–3; O. Neurath, 'Unified Science and Its Encyclopedia', [1937a], p. 270, *PP*, p. 164.
[83] O. Neurath, 'Physikalismus und Erkenntnisforschung', I, [1936f], *GpmS*, p. 754, *PP*, p. 164.
[84] O. Neurath, 'Unified Science and Its Encyclopedia', [1937a], p. 270, *PP*, p. 176.
[85] O. Neurath, *Le développement du Cercle de Vienne et l'avenir de l'empirisme logique*, [1936], p. 54.
[86] Ibid., loc. cit.
[87] O. Neurath, *Foundations*, [1944a], p. 25.
[88] O. Neurath, 'Einzelwissenschaften, Einheitswissenschaft, Pseudorationalismus', [1936a], *GpmS*, p. 705, *PP*, p. 134.
[89] O. Neurath, 'Soziologische Prognosen', *Erkenntnis*, [1936h], *6* (1936), p. 400; O. Neurath, *Einheitswissenschaft und Psychologie*, [1933], *GpmS*, p. 601.
[90] O. Neurath, 'Prognosen und Terminologie in Physik, Biologie, Soziologie', [1937b], *GpmS*, p. 788.
[91] Ibid., loc. cit.
[92] O. Neurath, 'Soziologische Prognosen', [1936h], p. 398.
[93] Ibid., pp. 398–9.
[94] The deductive-nomological model of explanation was formalised by C.G. Hempel and P. Oppenheim following Popper ('Studies in the Logic of Explanation', *Philosophy of Science*, *15*, 1948, pp. 135–75). See also C.G. Hempel, 'Deductive-Nomological vs. Statistical Explanation', in *Minnesota Studies in the Philosophy of Science*, ed. by H. Feigl and G. Maxwell, III, Minneapolis: The University of Minnesota Press, 1962.
[95] O. Neurath, *Foundations*, [1944a], p. 30 ("... many scientists prefer the above-mentioned analysis under the very indefinite assumption of *rebus sic stantibus*").
[96] O. Neurath, 'Prognosen und Terminologie in Physik, Biologie, Soziologie', [1937b], *GpmS*, p. 788.
[97] O. Neurath, 'Soziologische Prognosen', [1936h], p. 399.

[98] O. Neurath, 'Prognosen und Terminologie in Physik, Biologie, Soziologie', [1937b], *GpmS*, p. 788.
[99] O. Neurath, 'Soziologische Prognosen', [1936h], p. 399.
[100] Ibid., loc. cit.
[101] Ibid., loc. cit.
[102] Ibid., loc. cit.; O. Neurath, 'The Orchestration of the Sciences by the Encyclopedism of Logical Empiricism', [1945/46], p. 498, *PP*, p. 232; O. Neurath, 'Universal Jargon and Terminology', [1941], pp. 134–6, *PP*, pp. 218–20.
[103] O. Neurath, 'Soziologische Prognosen', [1936h], p. 399.
[104] Cf. C.G. Hempel, 'Deductive-Nomological vs. Statistical Explanation', passim.
[105] O. Neurath, 'Wege der wissenschaftlichen Weltauffassung', [1930/31], pp. 122–3, *PP*, p. 45; O. Neurath, 'Prognosen und Terminologie in Physik, Biologie, Soziologie', [1937b], *GpmS*, pp. 789, 793; O. Neurath, *Foundations*, [1944a], p. 24.
[106] Ibid., pp. 20, 30.
[107] O. Neurath, 'Soziologische Prognosen', [1936h], p. 399. Astronomers should wonder – Neurath adds – whether the degradation of uranium depends upon a cosmic situation which will change within some millions of years (ibid., pp. 399–400).
[108] Ibid., p. 404.
[109] Ibid., loc. cit.
[110] Ibid., loc. cit.
[111] Ibid., loc. cit.
[112] Cf. K.R. Popper, *The Poverty of Historicism*, London: Routledge and Kegan Paul, 1960, pp. 97–104. Popper polemically refers to p. 399 of Neurath's 'Soziologische Prognosen' (ibid., p. 103).
[113] K.R. Popper, *The Poverty of Historicism*, p. 103.
[114] Ibid., loc. cit.
[115] On Popper's anti-historicism see for instance: P. Winch, 'Popper and Scientific Method in the Social Sciences', in P.A. Schilpp (ed.), *The Philosophy of Karl Popper*, pp. 889–904; A. Donagan, 'Popper's Examination of Historicism', ibid., pp. 905–24.
[116] O. Neurath, 'Unified Science as Encyclopedic Integration', [1938d], p. 21.

CHAPTER 6

STRENGTHS AND WEAKNESSES OF AN EMPIRICAL SOCIOLOGY

1. LOGICAL EMPIRICISM AND THE SOCIAL SCIENCES: HEMPEL'S ANALYSIS

The relationship between logical positivism and the philosophy of the social sciences has received most detailed attention, with the exception of Neurath, from the neo-empiricist writer, Carl G. Hempel.[1] In his study 'Logical Positivism and the Social Sciences', an exploratory as well as historical examination of this relationship, Neurath's contribution is contrasted with that of Carnap and Feigl.[2]

According to Hempel's analysis it is possible to identify the following three main lines in Neurath's sociological thought:

1. Neurath displayed above all a strong 'materialistic' stance in his arguments against the 'idealistic' conception of science. He rejected categorically (even if, to Hempel's mind, with somewhat less rigorous arguments than those successfully employed by Carnap and other logical empiricists) the idealistic view of an essential difference between the historical and social disciplines (*Geisteswissenschaften*) and the natural sciences (*Naturwissenschaften*).[3] He refused to admit 'mentalistic' terms to the language of psychology and sociology, rejecting the idea that the subjective motives of the social agent could be conceived of as 'immaterial agencies'. He feared, as Gilbert Ryle was later to do in *The Concept of Mind*, that such a 'reification' of the individual conscience could lead to 'stultifying difficulties' in the relationship between immaterial 'mental agencies' and the physical world.[4] Neurath would therefore apply to sociology the general categories of Watsonian and Pavlovian behaviouristic psychology and would for the same reason be opposed to the sociology of Weber;[5]
2. In Hempel's view Neurath's political ideas and social concerns were basically motivated by Marxism, even if his adherence to it was not dogmatic but at all times open to discussion and critical reappraisal. Hempel maintains that in comparison with Carnap, who showed himself in all his writings to be a "highly rigorous and systematic"

thinker achieving "precise and explicit formulations of philosophical ideas", Neurath manifestly based his ideas and programmes on 'ideological considerations' without troubling to support them with analytical and systematic argumentation. According to Hempel his style was summary, elusive, and often "frustratingly vague on points of systematic detail";[6]

3. Neurath's appreciation of the role and potential of the social sciences – as indeed of any scientific enterprise – had in it strong pragmatic and instrumentalistic features. In Hempel's opinion he was not concerned with understanding or explaining the human world. For him the purpose of sociology, even when aiming to establish general laws of human change, was not to explain social phenomena, but solely to furnish the bases for empirical prediction of future events.[7]

Coming as it does from one of the most influential authorities on logical empiricism, this description of Neurath's sociological thought commands attention, and not simply on account of the claim it has to be an 'authentic' interpretation. It is also important in that it is written from the point of view of an 'orthodox' representative – even if more open-minded and flexible than most – of mid-European logical empiricism. It comes furthermore from the pen of one who brought to the centre of the English and American philosophical debate in the Forties and Fifties the problem of the application to historical and sociological explanation of the deductive-nomological model[8] (and in so doing made a decisive contribution to the formation of the 'standard view' of empiricism which was to exert so much influence in establishing the epistemological standing of the 'behavioural sciences' in the United States and Europe).[9]

Hempel's discussion is therefore of interest for the selective angle it applies to the examination of Neurath's sociological thought:[10] the interest lies as much in the areas it leaves out of account as in those it highlights for interpretation (and to a certain extent explicit evaluation) from the standpoint of the empiricist 'standard view'. We shall find it useful, therefore, in the following discussion of Neurath's political sociology, to keep closely to the guidelines provided by Hempel and to return at the end of the chapter to a critical review of his overall appraisal.

2. NEURATH'S CRITICISM OF GERMAN HISTORICISM AND THE PHILOSOPHY OF VALUES: MILL VERSUS DILTHEY AND MARX VERSUS WEBER

Even in his early work on Wilhelm Wundt's *Logik*, 'Zur Theorie der Sozialwissenschaften',[11] Neurath had cast doubt on Wundt's distinction between *Naturwissenschaften* and *Geisteswissenschaften* and had argued by reference to Karl Pearson's *The Grammar of Science* for the necessity of a wholesale reconsideration of the traditional partition between disciplines, including the notion itself of 'social sciences'.[12]

He did not return to the theme in any depth, however, before 1931, in two sociological works of seminal importance, *Empirische Soziologie*[13] and 'Soziologie im Physikalismus'.[14] Here the physicalistic proposal of an empirical sociology to be understood as *Sozialbehaviorismus* took as its starting point a well-defined critique of the metaphysical premises behind romantic historicism and the philosophy of values from Dilthey to Sombart, Rickert, Windelband, Spann and Scheler. The epistemological choice between social science and social metaphysics came down essentially to the two-fold choice between J.S. Mill and Dilthey on the one hand and Marx and Weber on the other, choices in which Marx was to be reinterpreted through the empiricism of Mill, and Weber understood principally through the aspects of continuity presented by his 'understanding' sociology (*verstehende Soziologie*) in comparison with Dilthey's historicism.

The opposition of Dilthey to Mill and of Weber to Marx was Neurath's means of showing that sociology as an empirical and historical science was the study of human behaviour as 'observable' and as amenable to being treated by means of empirically verifiable predictive procedures.[15] The conditions of 'observability' of behaviour and of 'testability' of predictions were procedurally satisfied by the analytical use of spatio-temporal coordinates, or, where this was not possible, by the use of other equally 'testable' categories.[16] This was obviously the case in physics and biology, but was also the case in psychology, sociology, and political economy. Differentiation of the methods of research in the natural sciences from those in the human sciences could be justified, he thought, only on the basis of the assumption that the objects of research were essentially different in the two cases; on the assumption, that is, that the behaviour of human subjects could not be the object of observation because it differed in certain crucial ways from every other type of behaviour ascribable to living beings or physical objects.

On this central theme, therefore, Neurath carried out a spirited and resolute assault against both the methodological theses of Dilthey's *Einleitung in die Geisteswissenschaft* and the philosophy of values of Rickert, Windelband, and Scheler.[17] He challenged the idea that a method peculiar to psychology and the social sciences could be identified in the processes of *Verstehen*, that is in the 'understanding' of the subjective motivations of social agents reached through an intuitive and sympathetic disposition (*Einfühlung*) on the part of the psychologist and sociologist.[18] He did not argue against the view that, in physics as in sociology, intuition may be a necessary condition of good research: if the term 'understanding' referred to the psychological gifts of a researcher, there was obviously no doubt that this element played a heuristic role in scientific research, just as it was obvious that a 'good cup of coffee' might make a scientist alert and stimulate his serendipity.[19] What he did deny was that the subjective motivations of the social agent could be conceived of as 'mental' or spiritual entities, impenetrable to empirical research and accessible *solely* to subjective intuition on the part of the researcher. Also he argued against the view that empirical research on the subjective motivations of the social agent could *take the place* of research on the regularities of observable social behaviour or prove itself to be more effective from the predictive point of view.[20]

Fred Dallmayr and Thomas McCarthy have charged Neurath with the formulation of a 'cup-of-coffee theory of understanding' which was rapidly taken up by other logical empiricists, such as Hempel, Oppenheim, and Nagel, and used as a 'standard weapon' in their conceptual arsenal.[21] In truth, however, as Georg H. von Wright recognised in his *Explanation and Understanding*, Neurath's arguments give him good cause to be seen on the contrary as the first sociologist to have formulated an accurate rebuttal of the neo-Kantian and neo-Hegelian criticisms of empirical sociology.[22] The fact that the question of *verstehende Soziologie* has been able to pass beyond its primitive, purely psychological, stage and to develop in a more profitable semantic direction is also owed to this rebuttal.

From Dilthey's historicist standpoint (developed in a spiritualist direction by Sombart, Windelband and Rickert), psychology, sociology, and historiography were, Neurath argued, as 'idiographical' or individualising disciplines, considered to be different from the 'nomothetic' or generalising sciences which investigated social regularities. The acquisition of knowledge, in the field of the 'sciences of the spirit', meant elevation to spiritual understanding of the world of 'freedom', 'inner experience', 'ought to be', or of 'values'.[23] Such positions naively reproduced the traditional essentialism and theological dualism by which human beings are aprioristically granted the privileged status of an entity totally different from any natural or biological phenomenon, and, in man, the intellectual faculties are hypostatised as 'spiritual nature', 'soul', or the 'noble' part concealed within its opposite, the body, and destined to survive its demise.[24]

Neurath did, however, acknowledge that, despite its encumbering metaphysical residue, German historicism contained in Dilthey and especially in Weber advanced methodological discussion and an impressive contribution to historiographical and sociological analysis.[25] The thesis maintained by Dilthey as well as by Weber, by which recourse to the 'understanding' of the subjective 'motivations' of social agents was necessary in order to guarantee the possibility of significant scientific discoveries, could not be simply dismissed as metaphysical.[26] In one sense it was the expression of real difficulties as yet not overcome by empirical research on social behaviour, and in another it provided positive information which needed translation into the language of scientific psychology and sociology.[27]

The scientific element which, he argued, was confusedly expressed by metaphysical terms such as 'soul', 'person', and existential '*Erlebnis*' should be capable of being reached through an empirical analysis of human behaviour which took into account also 'verbal behaviour'.[28] If a person gave a verbal description of the condition of being angry, then – he said, with some diminution and even contradiction of the force of physicalist behaviourism – it was certainly permissible to extrapolate from this verbal behaviour the 'organic perceptions' (*Organempfindungen*) which the observer knew from direct personal experience to be connected with the emotional condition of anger.[29] To this degree 'understanding' and 'empathy' could, he suggested, be seen as inductive processes as valid and as subject to restrictive conditions as induction was in the physical sciences. He acknowledged, however, that in these areas empirical psychology, even in Watson and Pavlov, operated with little terminological clarity and awareness, and that the relation between 'internal perceptions' and physicalistic propositions was expressible in quite complex forms not readily amenable to analysis.[30]

What the sociologist had to be at pains to avoid, he argued, was the use of expressions such as 'the nature of a people', 'the spiritual process', and – a phrase widely employed by Weber – 'the spirit of an age' (*Geist eines Zeitalters*)[31] if it was not totally clear that with these reference was intended to 'verbal combinations, religious forms, architectonic models, artistic styles etc.', i.e. to social phenomena and not to metaphysical substantiations or to spiritual entities understandable solely through sympathetic intuition. That the sociologist attempted to predict the behaviour of people in different periods by extrapolation from knowledge of his own behaviour is perfectly legitimate, even if at times misleading.[32]

But "the claim that 'empathy' possessed some kind of special magical property which made it possible to transcend the ordinary procedures of induction" was "unacceptable".[33] Equally untenable was the idea that 'classifications' operated solely in the sciences of nature while in the 'sciences of the spirit' understanding was also required.[34]

Empathetic understanding of people's motivations, aims, and desires could be useful and at times even necessary in the course of empirical research of social regularities and in the formulation of predictions. These aspects could not therefore be ignored by social science.[35] It was however

practical research on the behaviour of individuals and of groups which shows how much more reliably predictions are generally reached when based not on these elements of intention, attainable essentially through 'self-observation' *(Selbstbeobachtung)*, but on other elements of knowledge which can be acquired largely through other types of observation.[36]

This was the basis on which Neurath carried out in the course of a number of his writings a close examination of the methodological presuppositions underlying Weber's sociology. He made no attempt to hide the great respect in which he held Weber's work, a respect which was fully reciprocated in the many references Weber made in his *Wirtschaft und Gesellschaft* to Neurath's ideas on 'socialisation'.[37] Neurath credited Weber with having "attempted more than any other to make of sociology an empirical science"[38] and having "rendered a great service to modern sociology, especially through his work on the sociology of religion".[39] But Weber's methodology appeared to him to be flawed by a deep-seated ambiguity: in many ways Weber's sociology came across as a 'pure sociological science' but in many others, notably in his opposition on principle to behaviourism and his battle in support of a '*verstehende Soziologie*', he appeared to represent the continuation, via Windelband and Rickert, of Dilthey's spiritualistic historicism and methodological individualism.[40]

In Weber's famous *Roscher und Knies und die logischen Probleme der historischen Nationalökonomie*, as well as in *Wirtschaft und Gesellschaft*, Neurath detected a substantial trace of Dilthey's dualism between *Naturwissenschaften* and *Geisteswissenschaften*. Like Rickert before him in *Die Grenzen der naturwissenschaftlichen Begriffsbildung*, Weber tended to put 'empathetic insight' (*einfühlendes Sichversenken*) in place of the search for empirical correlations, and this opened the way to his use of conceptual categories for which no form of methodological test was conceivable, as with his concept of a 'rational economic ethos' (*rationales Wirtschaftsethos*), by which he meant a kind of social energy produced by Protestantism which would place science at the service of technology and the economy. This was the case also with the notion of 'Protestantism' itself, which Weber described as a 'force active upon people'.[41] He made a 'grandiose endeavour' to credit Calvinism with a form of causal pre-eminence in the definition of social behaviour in the modern capitalist world: from the 'spirit of Calvinism' had arisen the 'spirit of capitalism' and with it the capitalistic economic order.[42] But he appeared not to have addressed himself to the problem of the concealment of ideology traditionally carried out by preachers of Christianity, and especially of Catholicism, in the face of political and economic power, a point made very plain by study of the history of the papacy.[43] Even in the case of the relationship between Calvinistic ethical preaching and the productive organisation of capitalism, a Marxist interpretation, as supported by the findings of Catholic writers such as A. Fanfani[44] and above all the Jesuit J.B. Kraus,[45] seemed more cogent.

For, he argued, the doctrinal subtleties of the Calvinist theologians had largely passed the ordinary merchant and trader by and had not appreciably influenced his behaviour. It could more plausibly be claimed, for example, that the English merchants antagonised by royal monopolies and moneylenders who wanted an increase in the rate of interest approved by the Catholic church had been therefore disposed to support the doctrines of the Calvinist party which was opposed to the interests of church and crown. The behaviour of these men was already directed towards capitalism and only subsequently would they have been converted to Calvinism. And then, as a consequence of their allegiance, doctrinal formulations favourable to capitalism had appeared in Calvinism, a religion which originally had links with medieval pauperistic doctrines.[46]

The diffusion of particular religious doctrines and their emergence as motivating sources of human behaviour could be seen, then, not as a causal factor, but as a variable dependent on surrounding conditions, and it became a matter therefore of discovering which circumstances had formed the collocation (*Zusammenwirkung*) which caused the new cultural attitudes, motives, and modes of behaviour to take the place of the old ones. In any given situation, in other words, the driving forces of social change had to be identified.[47] Weber, however, despite the depth of his historiographical work, had prevented himself from grasping this central issue by his aprioristic assumption of an 'empathetic' approach.[48]

The methodologically correct solution to this problem, Neurath argued, required analysis of the many and complex variables present in the general social conditions of the group under consideration and then the attempt to establish empirical correlations between these general conditions and the particular modes of behaviour whose appearance it was intended to identify or predict.[49] It was this attempt which distinguished Marxist empirical sociology and set it in opposition to the 'idiographical' standpoint of *verstehende Soziologie.*

3. MARXISM AS EMPIRICAL POLITICAL SOCIOLOGY

Neurath's social philosophy was originally formed under the influence not so much of Marxism as of that particular type of political utopianism and social positivism which, in Viennese enlightenment of the early years of the century, found its most notable interpreters in Josef Popper-Lynkeus[50] and Carl Ballod-Atlanticus,[51] and, in Germany, in the war economist and 'scientific utopian', Walter Rathenau.[52]

His early thought on wartime economic planning and the possibility of adoption in peace also of a planned economic model based on the widespread and systematic use of social statistics was directly influenced by the schemes of social engineering developed by Ballod-Atlanticus and Popper-Lynkeus.[53] The doctrine of 'socialisation' which lay behind his actions in 1919 as Director of the Office of Central Planning of the Bavarian Republic, was built on essentially non-Marxist lines.[54] In practice his greater involvement with Marxism came only after the first World War and especially when he began collaboration on *Der Kampf*, with full adherence becoming evident only some years later. In fact Marxism as a term of reference in Neurath meant much less classical Marxist thought than the more direct influence of the Austro-Marxist version represented by such writers as Max Adler, Otto Bauer, Rudolf Hilferding, and Karl Renner.[55] Neurath's Marxism was therefore characterised by the repudiation of Hegelianism and dialectical materialism, by explicit reservations on Lenin's interpretation of empiriocriticism,[56] and a wide-reaching attempt to harmonise Marxist doctrine with modern scientific thinking, particularly with Mach and Avenarius. It was a form of Marxism which, like Max Adler's, was especially concerned with the theory of knowledge, philosophy of science, and the methodology of social research.[57]

It is in *Lebensgestaltung und Klassenkampf* of 1928 and in the two sociological writings of 1931 that we find Neurath's original interpretation of Marxism reaching a developed state.[58] In order to counteract the claims of Weber's *verstehende Soziologie* he portrayed Marxism as the only sociology able to meet the methodological requirements of physicalism and 'social behaviourism' (*Sozialbehaviorismus*).[59] While he acknowledged that there were 'numerous metaphysical tendencies' in Marx and especially in the Marxist tradition, that Marxist analysis of capitalistic societies was based on an excessively simplified outline, and that the very formulation of the relation between 'base' and 'superstructure' presented unresolved problems,[60] these did not invalidate for him the empirical nature of Marxist analysis which remained the most powerful methodological alternative to the spiritualistic impostures of the philosophy of values and the methodological individualism of *verstehende Soziologie*.

The merit of Marxism, as Neurath saw it, was that it furnished a conceptual frame which, while not ignoring the subjective 'ideological' motivations of social agents, as expressed by their verbal behaviour, still managed to avoid taking them as 'mental' premises of sociological explanation and prediction. In the Marxist system the search was not for "analysis of the 'motivations' of individuals as if this could provide a sufficient platform for the identification of relations formulable as laws".[61] People's assertions about the 'reasons' for their conduct were "fundamentally more dependent on contingent factors than would tend to be the case with their more general behaviour".[62]

If the social conditions of a certain historical period were known,

> we can predict the behaviour of whole groups much more readily than the rational justification which individuals will give of their conduct. Individuals justify their conduct in quite different ways. Besides, few people notice the connection between the social situation and average behaviour.[63]

There were convincing examples of this to be found, he observed, in the decline of slavery in the ancient world and in the emancipation of slaves at the time of the American civil war. In the Roman period, as the agricultural writer Columella informed us, slaves were progressively freed for the reason that their cost was economically disadvantageous in comparison with the use of unenslaved workers provided by the market at any given time, for whose illness or death the provider of work carried no responsibility.[64] Furthermore the idea that the spread of Christianity had been the principal cause of the disappearance of slavery was contradicted by the fact that the most oppressive forms of slavery had appeared just when the Christian countries reached their greatest point of expansion and when the Catholic church exercised its most powerful influence on the colonial territories. It was no accident that the importation in cargo ships of black slaves to Spanish America met with the favour of Catholic theologians, since they wished to preserve the Indian slaves, at that point in danger of extinction.[65] In North America emancipation of the slaves inflicted serious losses on the cotton producing states of the South, but not on the industrial Northern states, which was highly likely to have been a reason for the civil war. On the other hand very little light seemed likely to be shed by an inquiry into the precise 'motivations' of the slave-owners in Rome, the Spanish colonies, or North America who freed their slaves: few would recognise that they had done so because of the insufficient return yielded by slavery; instead, humanitarian or religious motivations would be claimed by the majority.[66]

An empirical sociology based on Marxism would attempt rather to establish correlations between the behaviour of classes and their social situation: in this way it would attempt to provide an explanation on the basis of observable regularities for the often variable 'verbal motivations' which the idiographical approach took to be 'motives' of behaviour.[67] In this specific sense Marxism could be assimilated to those schools of psychology, such as psychoanalysis and 'individual psychology', which attributed an important function in human affairs to the unconscious and did not aprioristically aggrandise 'conscience' to be the seat of human 'free will'.[68]

What was the nature, however, of the correlations and observable regularities which Marxist sociological empiricism saw as important for establishing social prediction? In order to predict with any degree of credibility the future behaviour of a social group it was "necessary to know the whole life of the group". The correlations and interdependencies of the various elements in social life were so numerous that it was impossible to form predictions on aspects of technology, religion, or everyday habits, while isolating them from their context or taking no account of strict temporal limitations. In order to predict, for example, the ways in which expressions connected with religious life (e.g. the structure of religious buildings) would change within a social group, it was necessary to consider "the mode of production (*Produktionsweise*), the forms of social organisation, and the types of religious behaviour" current at that fixed historical moment and to discover "the transformations to which *all these elements together* are subject".[69]

Neurath saw the fundamental methodological hypothesis of Marxism as consisting in the relation of asymmetry which it introduced into the web of interconnections linking the multifarious factors of social change. Given certain conditions, it was possible, according to the Marxist hypothesis, to infer approximately from the structure of the 'mode of production' within a certain social group not only the future stages of the development of its productive organisation but also the general formation of its social organisation (*Gesellschaftsordnung*), and on the basis of this prediction further hypotheses could be advanced with some chance of success on the future religious behaviour, architectural style, musical culture etc. typical of that social group.[70] But "experience shows that the reverse process is not valid": it was not possible to draw predictions concerning the mode of production from hypotheses concerned only with the element of religious behaviour.[71]

Through the adoption, he argued, of this interpretation based on the asymmetry between one set of social structures or 'basis' (*Unterbau*) and another, the 'superstructure' (*Überbau*), Marxism did not create any conflict between a 'material' and a 'spiritual' sphere or introduce different types of causality, each ascribable to the different 'essences' of 'mind' and 'matter'.[72] Marxism was concerned with the search for correlations and empirical regularities which could take the form of sociological laws of social change: such laws would be of a general nature or, in very many cases, would possess a validity limited in time and space.[73] Most of the regularities in sociology in fact held only for restricted periods of time and within strictly defined limits, as was the case also in biology, where, alongside a small number of general laws, there were particular laws relating to specific species of animals; those sociologists were therefore in error who held that the laws of their subject should have the same nature as the laws of chemistry and so be valid under all the circumstances conceivable on our planet.[74]

An empirical sociology based on Marxism recognised, then, that "in sociology the general laws are very few in number"[75] and united analytical research on the components of social change with historical awareness of the contingent nature of the social systems which actually arise.[76] Marxism was prevented by its 'critical and scientific attitude' from believing in the possibility of prediction over a long period:[77] only 'highly approximate indications of historical tendencies'[78] were scientifically possible. The Marxist sociologist was well aware how very limited the scope was for the kind of knowledge of historical conditions which was necessary for long-range prediction,[79] also that the course of history was always open to differing interpretations. Practical attempts to reach long-term objectives as well as the very historical and sociological knowledge on which the social scientist – the Marxist scientist included – relied in order to make his predictions were conditional on complex historical circumstances.[80]

Above all Marxism applied to itself the Marxist doctrine of the structural asymmetry of the factors of social change and consequently of the historical and sociological conditioning of knowledge.[81] Social predictions were elements of scientific theories: as such they were also "sociological facts fundamentally dependent on the social and economic order".[82] On the other hand sociological inquiry was both a stimulus to and an instrument of social organisation, and this led to the close connection of the growth of sociology with social conflict. Only established schools of sociology which had dispensed with social support and financing could use their collective work to master the large quantity of material necessary for the correct development of hypotheses and social predictions.[83] Sociological research could not expect to be any less 'socially neutral' than other sciences.[84] Marxist empirical sociology saw sociological research as one social *praxis* among others: the credibility of results produced by any given scientific research depended less on the private aspirations of the individual researchers than on the social position of those permitting, supporting, or financing the research.[85]

Thus it was possible to maintain that the difficulties which beset Marxist empirical sociology were caused less by any theoretical defects than by the opposition of social groups who saw it – not without reason – as a threat to doctrines important to the maintenance of the prevailing order.[86] All the same, Neurath declared himself far from content with the sociology of knowledge elaborated by the 'bourgeois Marxist' Karl Mannheim in *Ideologie und Utopie*. Mannheim saw every ideology as a kind of distortion or partial consideration of reality, as a 'unilateral' point of view in relation to others, and in essence no different or better than the rest.[87] Neurath subscribed to Mannheim's ideological relativism, but was critical of what he saw as the contradictory attempt it made to use the sociology of knowledge in order to reach a way of thought liberated from ideological presuppositions. There did not exist, he objected, any neutral and higher standpoint, any all-embracing *Weltanschauung* with claim to be a rational synthesis of the 'unilateralities' of ideologies, or a 'total truth', or an 'omnilateral method', a height from which to look down on ideologies as partial doctrines. Mannheim's *Wissenssoziologie* claimed in various ways to transcend the self-establishing reflexivity of every form of knowledge. In practice, however, knowledge, whatever form it took in the science of nature as of man, remained inescapably unpossessed of fixed and absolute foundations. The extent of its scientific nature could be measured only by the degree of credibility contained in its empirical predictions. From this point of view too Marxism was a science and not a metaphysical *Weltanschauung*.[88]

4. SOCIOLOGICAL 'PSEUDORATIONALISM': THE INADEQUACY OF BEHAVIOURISM AND THE 'OVERMATHEMATISATION' OF SOCIOLOGY

Foundations of the Social Sciences, which was published in Chicago in 1944 in the series *Foundations of the Unity of Science*, formed a premature conclusion to Neurath's scientific work and a final testimony to the strength and fruitfulness of his philosophical thought.[89] Here he subjected social behaviourism to the scrutiny of a severe criticism which was also to a large extent implicit self-criticism. No longer was the assault on the metaphysical assumptions of romantic historicism and the philosophy of values conducted in terms of a strict opposition between the method of *verstehende Soziologie* and the predictive nomology of social behaviourism: instead inquiry into the bounds of sociological nomography, the inevitably conventional and pluralistic nature of social understanding, and the conditioning, difficulties, and paradoxes behind social prediction were now brought to the fore. Criticism of Weber was no longer directed against his 'mentalism' but against the conceptual problems associated with the formulation of 'ideal types' of social behaviour[90] and, implicitly, against the vestiges of a causalist conception of sociological explanation contained in the use of asymmetrical correlations between the factors of social change. For this same reason, as we shall see, the interpretation of Marxism as empirical sociology was in one important respect altered and in fact abandoned.

In *Foundations of the Social Sciences* Neurath strongly reiterated the empiricist requirement: sociological language should be free to the greatest possible extent of propositions not linked or capable of being linked to observation sentences. He granted, however, that metaphorical expressions such as '*Volksgeist*', '*Ethos einer Religion*', '*etische Kräfte*', widely employed by Weber and 'oceanic feeling' used in psychoanalysis could not be suppressed simply because they were judged to be 'metaphysical'. Such suppression would result in no more than a pointless impoverishment of sociological language, since empiricist reduction of such social or psychological metaphors proved in practice to be highly difficult and complex.[91]

Sociology, like any other science, including physics, took as its point of departure not the (supposed) precision and univocality of the so-called 'sense data' (or 'simple basic assertions' or 'atomic ideas'), but the "full lump of irregularities and indistinctness, as our daily speech offers it"[92] from which even 'observation language' could not be freed. In fact the empiricist

> cannot go on without handling 'indistinctness' everywhere, since he wants to realize the program of assaying assertions by means of observation-statements. Observation-statements cannot but be indistinct in some way or another.[93]

In the social sciences assumption of the presumed clarity and determinacy of the 'sense data' as point of departure would entail a kind of 'scientific suicide', since sociological research not only could not relate to anything 'elementary' or 'original' but required a great richness and indistinctness of expression.[94]

Moreover, in the field of the social sciences and historiography, scientists needed to be able to describe not just the external behaviour of social agents but also their 'emotional state', 'internal discourse', and feelings of faith, hope, and fear.[95] From this point of view not only were the deficiencies of Watson's behaviouristic psychology revealed,[96] but also the poverty of any behaviouristic terminology became manifest:[97] behaviourism did not succeed in expressing the emotions aroused by art, literature, religion, or feelings of the heart, for which a delicate and complex language was needed.[98]

Thus, for example, though it was possible to point out a limitation in Weber's treatment of the relations between 'capitalism' and 'Calvinism', this did not affect the wholly acceptable hypothesis that important sociological correlations could be established between these 'historically given items'. The limitation was to be seen in the difficulties Weber had encountered in handling scientifically and with a sufficiently adaptable language not just the external behaviour of social agents but also their emotional states ('feeling-tone'). It was precisely this lack of a language capable of expressing emotional as well as behavioural states which restricted the productiveness of Weber's 'fine analyses' of the 'wonderful material' of the origin of capitalism.[99]

The pitfalls of social research were not associated solely, however, with the inadequate development of theoretical categories and linguistic constructs capable of expressing the objects of the research. Equally serious were the difficulties which arose from the extension into the areas of sociology and historiography of the 'pseudorationalistic' attitude so widespread in the natural sciences.

What sociological 'pseudorationalism' did was to deny the pluralistic, relative, and conventional nature of all social knowledge. It denied it above all when it extended an 'ontological' conception of scientific understanding to sociology and applied to social phenomena an absolutistic and logicistic notion of truth, such as that developed by Tarski and Carnap in the course of their 'semantic' research.[100] In sociology, Neurath argued, as in all other empirical sciences, an observation sentence could not be called 'true' (or 'false') *if and only if* a 'fact-datum' corresponded (or did not correspond) to it. Such an 'absolutistic' mode of expression could rightly be termed 'ontological', since it followed the ontological tradition going back to Aristotle.[101] In this system 'speaker', 'speech', and 'object' were treated as if they were three different players in a game, while a rigorously 'terminological' approach, assuming the 'non-transcendability' of language, implied that we must

> not start any discussions and meditations with ontological assertions but with the comparison of statements with statements and their terminological analysis.[102]

According to this approach – which was wrongly taken by its critics to be a 'coherence theory of truth' and was taken instead by Dewey and Bentley in its more precise sense[103] – an assertion was accepted or rejected on the basis of grammatically complex procedures which only ever concerned relations between propositions or relations between propositions and theories.[104]

When historical questions or questions concerning the gauging of sociological correlations were under consideration, Neurath argued in explicit disagreement with Tarski and Popper, not only should we avoid using the binary scheme of true/false and speaking absolutely of 'errors', but it was similarly out of place to refer to the 'truth' in an asymptotic sense, as if the 'truth' could be seen as a kind of mathematical limit "which cannot be reached completely but only to some degree". All that could be said about historical and sociological assertions was – as with all other observation sentences – that they become accepted by social scientists "at a certain time and at a certain place".[105]

In assessing the 'truth' of an observation sentence, Neurath claimed with consistent conventionalism, the historian and the sociologist found themselves in the position of a judge **without the power** to decide authoritatively which of two mutually exclusive pieces of evidence should be taken as true and so form the basis for his judgment:

> How may scientists (all being historians to a certain degree) discuss observation-statements made by eyewitnesses? What can they do when the statements seem to be incompatible? You may think, for instance, of judicial proceedings. The eyewitness

may say, 'The dress was gray'. You may be on the bench and order the beadle to present the dress, by saying, 'I want to discover whether the statement is true or false by observing the color of the dress'. And then you may say, 'After looking at the dress, I can declare the eyewitness' statement is not true: the dress is blue'. If you and the court have the authority to decide what be legally true, the matter is decided; in the republic of science other customs are valid. There is no such authority, and the eyewitness may send you into the witness box and ask questions from the bench and decide together with the court that your saying is false.[106]

Thus the republic of science possessed no authority or judge to decide absolutely whether an empirical proposition was 'true' or 'false', if by this was meant its 'correspondence' to reality, since not even the scientist's direct sense perception could establish definitively which assertion 'corresponded' to the data of fact.[107] What operated therefore in science was a general rule for the conventional and provisional acceptance of the methodological decisions made by researchers.

"There is no means" as he forcefully put it in *Prediction and Induction*, "for the definitive verification or rebuttal of a hypothesis through any scientific procedure whatever". Of course it was possible to reduce the multiplicity of scientific hypotheses at random or by declaring the judgment of one privileged person to be decisive. But these were not 'scientific means'. An empiricist sociology could therefore not fail to be irreducibly pluralistic. While

others speak of 'facts' or the 'truth' of something that is 'objective' and 'absolute', the logical empiricists speak of protocol sentences as a basis for further argument. We go far if we succeed in accepting some protocol statements as a 'common basis' of argument: in many areas it is difficult to achieve even this minimum of accord.[108]

It was in many ways dangerous to speak of 'truth' or 'reality', because these linguistic expressions appeared to allude to the existence of some kind of definitive sentence. The very expressions 'rule of reason' and 'rule of arguing' were to a certain extent contradictory, since arguing contained the indefinitive implication of something which could be argued.

In practice the social scientist always found in the end that, once he had rejected obviously unbelievable data and hypotheses, he was in the position not of having to decide between the 'truth' and the 'falsehood' of individual empirical propositions, but of having to choose between a broad range of competing propositions, all in some measure both plausible and acceptable. The determining element in this kind of choice was, as Neurath saw it, of an external nature, extrinsic to the field of methodology: it was a question of considering whether the researcher had at his disposal the necessary time and energy to test all the competing series of propositions or whether he should decide to work on just one without making investigations in other directions.[109]

In accordance with this strictly pluralistic and conventionalistic stance, Neurath maintained that a historian could write

> more than one history of the Thirty Years' War: more than one biography of Galileo, Wallenstein, or Cromwell, selecting sometimes these, sometimes those, acceptable and plausible statements.[110]

One could hardly, he argued, regard as scientifically serious a historian who denied the irreducibile nature of this pluralism and who argued that it was possible to reach an 'optimal' solution. Pluralism was a feature even of physics, as Duhem, amongst others, had shown.[111] Rather than speak, as the rationalists did, of the 'world system' or the 'universe', we should instead use terms such as 'pluri-verse' (*Pluriversum*), 'pluri-moon', 'pluri-table', 'pluri-Newton'.[112] In the historical and social sciences "this pluralism is more conspicuous, because relatively simple stories are full of hypotheses".[113] For this reason the tendency to introduce the category of 'optimisation' into the social sciences was to be resisted, with the sole exception of cases in which the 'optimal' reckoning belonged to a formal, rigorously defined, scheme. And this was true also, more generally, for the application of quantitative methods to sociological research.

Beyond doubt mathematical analysis of social correlations was attractive and stimulating, but, he argued, it was necessary to avoid the tendency towards 'overmathematisation' of the social sciences.[114] When quantitative reckoning related to items which required complex conceptual determination, as was frequently the case in sociology, even the most sophisticated logical scheme proved itself of little use from the practical point of view. Cases occurred, for example, where mathematical calculation proved itself irrelevant for the purposes of prediction, because it formalised aspects which could not be correlated with the sum of the other, sociologically relevant, variables.[115] For instance mathematical formalism in the Paretian and marginalist developments of price theory certainly provided ingenious and logically consistent conceptual constructions. But these showed themselves to be of little use if one tried to apply them to complex economic and social situations. It was no accident that Pareto and the marginalist school moved suddenly from mathematical rigour to poetical vagueness when they confronted the problem of the recurring crises of modern capitalistic economies.[116]

More often there were cases where the advance 'ranking' of concepts proved wholly problematical, and so too the selection and definition of the characteristic marks of the social phenomenon to which formalised calculation was to be applied. Because 'index figures' could be regarded as important and reliable, it followed naturally that the general theory was also important and reliable, by which the qualitative 'items' or marks were selected and defined, subsequently being treated mathematically and thus arranged in series of ordinal numerals.[117]

There was a misleading tendency therefore – to be seen for instance in Sorokin's *Sociology of Revolution* – to move from the ordinal gradation, on the basis of established conventional scales, of the different characteristics of a phenomenon (e.g. the intensity and duration of wars or revolutions) or of a social group (e.g. rates of mortality, suicide, illiteracy, spread of mass communication, national wealth) to calculation in cardinal numbers. By aggregation and combination of the (cardinal) index figures of the various scales one arrived at a unidimensional classification of the phenomena or social groups under consideration.[118] The arbitrary nature of these hyperempiricist procedures revealed itself, he pointed out with a reference to Bortkiewicz, when the numerical data were disassembled and it became clear that the final result depended, for instance, on the 'standard population' which had been taken initially as the very basis of the classification of the concepts and of the definition of the characteristics to be taken into consideration. For example, the death-rate in different social groups could not be expressed in cardinal figures without the advance assumption that the age distribution of the people in each group corresponded to the one conventionally allowed for in the 'standard population'.[119] Modification of this parameter of reference would result in modifications to the final results.

We had therefore to ask ourselves, he maintained, whether it would not be better to use in such cases 'sociological outlines' (*silhouettes*) which would record the ordinal valuations of the different characteristics under consideration and would restrict themselves to a more sensible multidimensional comparison between these characteristics. This would avoid producing the illusion (or self-deception) that the cardinal figures expressed precise and homogeneous values.[120] When it was matter of dealing, as it almost always was in sociology, with objects which were complex and difficult to analyse (*clots, Ballungen*), the use of a consciously imprecise language was preferable to highly sophisticated mathematical techniques.

The apparent univocality and precision of measurements and calculations did, in any case, leave the pluralism of social theories and the indeterminacy of sociological language entirely out of account. This meant that the scope was wholly unrestricted for those decisions and agreements which alone made possible the acceptance or rejection of the assumptions which necessarily preceded all types of mathematical formalisation:

we cannot without particular assumptions expect that the correlations within the calculus in which a true-false distinction (related to a certain language) appears, may be regarded as correlations between items in our empiricist field. [...] We cannot say within empiricism, 'Exact measurement teaches us...'. In the social sciences, as in other sciences, we often meet collected results of measurement, and then we try to substitute such collections (called 'belonging to the same item') by certain 'index figures', often called the 'correct' measures, whereas the others often are regarded as 'errors'. Sometimes these terms are harmless, but sometimes they lead into a 'truth' phraseology and support an absolutist viewpoint. In the formulas of calculatory schemes we may speak of 'exact measurement' and therefore of 'error' also. But we cannot apply the results of the calculatory schemes to the aggregational discussions without making certain assumptions.[121]

These considerations were sufficient to exclude from the social sciences both Popper's proposed falsificationist method and Carnap's theory of 'degrees of confirmation'. As we have seen, Neurath found several differing 'agreements' always to be possible in the determination of quantitative indices as well as in the inductive construction of statistical regularities in relation to different classes of reference. In fact

supporting an 'induction' implies producing uniformities, already given, combined with uniformities in 'trend', etc., for producing possible predictions. Of course, in harmony with logical empiricism we ask for the elimination of contradictions where possible and for the collecting of uniformities; but, finally, a margin remains for our decision again and again.[122]

Here more than ever became evident the inadequacy of Popper's logical scheme which attributed to physical or social regularities the form of 'universal laws' and which rested on the logical and methodological asymmetry between 'verification' and 'falsification'. If some scholars, Neurath argued, prefer

to work within a calculatory sheme of some 'universal laws' of the type 'Every A has the behavior *m*', they may say that any 'universal law' may be destroyed definitely by any contradicting hypothesis which seems to show that 'this single A has not the behavior *m*'. They may add that no 'universal law' could be 'verified' by empiricist statements, because that would need an infinite number of 'good' statements. This calculatory scheme, with an asymmetry very attractive to many scholars, is not valid for aggregational arguments which include all sociological sciences. In the field of aggregational analysis we always have to deal with more than one possibility. There is more than one convention possible in making index figures, more than one in making 'supplementations'.[123]

In sociology, more than in any other field, the researcher was faced with a complex web of hypotheses so interconnected that, if a prediction turned out not to be confirmed by experiment, it was impossible – Neurath observed, referring once more to Duhem – to identify which single hypothesis had been the cause of the failure: "we cannot say from which hypotheses certain difficulties arise".[124] Once again therefore it was necessary therefore to start from a pluralistic approach and to recognise the perfect symmetry between corroboration and 'shaking' (*Erschütterung*) of sociological hypotheses, for

'supporting' hypotheses by means of material suitable for hypotheses is based on decisions: we are selecting material; 'corroborating' hypotheses and 'shaking' them by means of 'positive instances' and 'negative instances' implies decisions; but no '*experimentum crucis*' may invalidate any single hypothesis.[125]

It was also necessary to recognise that

the 'positive instances' play a great role and not only the 'negative instances', as in the asymmetric refutation-verification scheme. [...] In the social sciences positive results form the backbone of success.[126]

Nor was it possible to maintain, as Reichenbach had thought, that the confirmation or disconfirmation by experiment of individual sociological hypotheses could be conceived of as distinct and enumerable *items* and consequently arrangeable in series or 'degrees of confirmation', as Carnap proposed in *Testability and Meaning* in his (unsufficient) attempt to liberalise the criterion of verification of the sentences. Every single 'confirmation' or 'disconfirmation', he said in conclusion, always belonged to complex 'aggregates of arguments', to which it was not possible to apply a scheme of formalised calculation. We may 'shake' or 'corroborate' the assertion of a hypothesis and finally prefer some hypothesis, but we would not be able to "confirm a sentence more and more".[127]

5. CAUSAL ASYMMETRY AND THE *CETERIS PARIBUS* CLAUSE IN SOCIOLOGY: THE LIMITATIONS OF FUNCTIONALISM AND MARXISM

Neurath identified a second and no less serious danger threatening the social sciences in the use of 'asymmetrical' schemes in anthropological and sociological explanation and especially in the adoption of causal asymmetry, falsely considered by social scientists to be the universal model of scientific explanation.[128]

This danger was present above all in functionalist anthropology, which used the term 'function' with asymmetrical meanings oscillating between the teleological sense borrowed from biology and the more properly causal one derived from physics.[129] But it was present also in the bio-ecological and socio-ecological theories which dealt with the antithesis between an individual and his environment, aprioristically assuming a causal asymmetry in favour of one or the other rather than representing the interaction between social groups and their environments in the form of a 'synusia' and so of a unitary aggregation typified by a symbiotic continuity of structural relations between physical, biological, and social phenomena.[130] It was especially present also in Marxism, whose basic explanatory scheme of structure/superstructure was a typical instance of sociological causalism.[131]

The use of a 'cause-effect phraseology' referred back to the idea that certain sequences of phenomena would recur "whenever the same conditions are given".[132] The causal explanation, in other words, introduced a tacit *ceteris paribus* clause as a condition of the recurrence of any given phenomenon: thus, as Mach and Frank had argued, causalism was either unaware of or disguised the fact that all phenomena were of necessity involved in highly complex 'cosmic aggregations' with uncertain boundaries and were by no means temporally invariable.[133] It was not without purpose that Mach had called on any scientist engaged in the causal explanation of phenomena at least to provide detailed identification of all the factors from which his experiment could not be isolated.[134] In fact, Neurath acutely observed, no practical scientist was in the position to rule out the possibility that his explanatory or predictive hypotheses might be weakened by experiment. In physics, as in the social and historical sciences, no phenomenon or element could in practice be isolated from the context of a particular 'cosmic aggregation' and removed from the influence of its temporal variations.[135]

In sociology and economics adoption of causal schematism had led in particular to the simplification typical of monofactorial theories.[136] Working on the basis of an 'artificial uniformity' obtained through implicit or explicit use of *ceteris paribus* (or *rebus sic stantibus*) clauses, some sociologists succeeded in establishing precise correlations. But causal analysis was normally able to establish the correlations of only a highly restricted number of elements in comparison with the very great complexity of social factors.[137] This was the reason for the remarkable frequency with which sociologists and economists had recourse to *rebus sic stantibus* clauses of an entirely indefinite nature. And it was the reason also why highly accurate and sophisticated studies – for example of market forces – possessed an empirically limited usefulness, since the general conditions could not be estabished under which the relations being studied would continue to hold true.[138] The insertion of the *rebus sic stantibus* clause into his causal explanation enabled the researcher to pass over exactly those *res* which a sociologist conscious of the interdependence and complexity of social phenomena could not leave out of account.[139]

Neurath therefore made the somewhat surprising suggestion[140] that sociologists, economists, and historians should abandon their claim to discover the 'causes' of social phenomena: it would be more useful if they were to limit themselves to showing how certain events or conditions were in the process of 'emerging from' or 'growing out of' certain social aggregations. The task of the social scientist was not to construct *a priori* laws and general relations of an asymmetrical nature[141] but to evaluate in each individual case the kind of connection which linked certain phenomena to other *particular* phenomena.

This was generally true also for the Marxist hypothesis of the dependence of 'superstructural' phenomena on the economic 'basis' of a certain 'mode of production'. Apart from the difficulty of accepting the thesis of the causal priority of the economic 'basis' – modification of sociological theses, for instance, was not a simple 'symptom' but also a 'co-conditioning factor' *(mitbedingender Faktor)* of social and economic transformation[142] – it was hard even to attempt an empirical version of this most general of Marxist assumptions. The most one could do was to set out to investigate as a *quaestio facti* whether the explanatory asymmetry generally claimed by Marxism translated itself in individual cases into a 'predictive asymmetry' *(Prognosenasymmetrie)*.[143]

It was a question, in other words, of ascertaining whether, by taking as an empirical basis the expressions of art or the forms of legal organisation or, more generally, the theoretical formulations (i.e. opinions, projected schemes, oral or written verbalisations) which expressed the nature of a social group at a given historical moment, it was possible to explain the forms of that group's productive organisation and to make predictions about its future developments.[144] Or whether instead only the reverse process was possible, to explain the group's expressions of art or its forms of legal organisation by starting from its economic organisation and basing predictions on that.

And this, Neurath claimed, even though Marxist theorists would only accept such an empirical transformation of historical materialism with great difficulty,[145] was the sole 'modest scientific residuum' (*bescheidenes Überbleibsel*) of the 'asymmetry terminology' which characterised the Marxist doctrine of the relation between 'basis' and 'superstructure'.[146]

6. PROBLEMS AND PARADOXES IN SOCIAL PREDICTION: THE ROLE OF REFLEXIVITY

The central aim of sociology as an empirical science was, in Neurath's conception, the prediction of social change.[147] It was a nomographic discipline whose field of study was society's 'structure', internal connections, and external interactions with the environment.[148] The task of the empirical sociologist was to develop theoretical hypotheses which would allow him to understand and predict the behaviour of social groups on the basis of their past history. He worked, that is, by isolating first, as far as he could, the basic conditions of these groups through statistical recording of regularities in their behaviour and by the systematic employment of the comparative technique.

It had been well shown, he maintained, by Machiavelli and Montesquieu[149] that comparative political historiography in particular had much to teach us about the understanding and prediction of social behaviour. The achievements of these founders of the empirical approach to political science had been to show the relative invariability of political behaviour (and of political systems) despite the variability of historical and geographical conditions, also to rid the writing of political history of underlying moral asssumptions or projections of divine involvement, achievements which were conspicuously absent in Hegel's 'gigantic theodicy' (*gigantische Theodizee*).[150]

The contribution of empirical sociology would become much more important, however, if work on political history would ally itself to the kind of research in *integrated* anthropology, historical economics and political economics, which took as its central theme the living standards of individuals or of social groups and addressed itself to problems of their empirical definition, social distribution and stratification, and their variations in harness with the fluctuations of other political and economic factors.[151]

Neurath's expectation was that sociology practised in this way would not only produce significant theoretical results but might also provide bases of knowledge and means of prediction capable of bringing about a rational and systematic restructuring of society. Sociological prediction, based on a 'theory of the social structure' was inseparable, to his mind, from political planning: social engineering required the constant employment of sociological theories, while the practical application of these theories required continuous theoretical correction and reformulation. Political activity practised in the 'grand style' (*großen Stils*) – we note the contrast with Popper's 'piecemeal tinkering'[152] – was the most effective means of providing experimental testing of sociological theories. Close connection could thus be said to exist between the predictive side of sociology and political planning understood as 'live social engineering' (*lebendige Gesellschaftstechnik*). It could even be said that the very chances of success of sociological theories were intimately bound up with the conscious restructuring of society.[153]

Neurath's confidence in the capability of 'integrated' social science to provide the basis of a rational planning of political and social development was, however, greatly counterbalanced by considerations of an epistemological kind. Sociological prediction was, he recognised, subject to limitations, problems, and paradoxes over and above those which generally applied to all types of empirical prediction.[154]

Leaving out of account their capacity to draw more widely on the experience of the past, sociologists "are no better off than the man in the street as far as their predictions are concerned".[155] As he wrote in one of the final pages of *Foundations of the Social Sciences*

> as social scientists, we have to expect gulfs and gaps everywhere, together with unpredictability, incompleteness, and one-sidedness of our arguing, wherever we may start.[156]

In the empirical sciences and so too in the social sciences – it had to be admitted straightaway – hypotheses, and the predictions based on them, were no more than 'working hypotheses', if the euphemism of such 'consoling phraseology' was really to be used.[157] In practice, although the procedures of all empirical sciences were essentially identical – the difference being only of degree or of a technical nature – it was necessary all the same to recognise that especially difficult problems were encountered in the field of social sciences. It was here that prediction ran into the greatest difficulties.[158]

Above all it was essential to keep in mind the 'great variety of possible changes' (*die große Fülle möglicher Veränderungen*) which were typical of the evolution of human groups, their political institutions, and their organisational structures. This evolution was of a particularly complex and therefore unpredictable kind.[159] The reactions of individuals and groups to exactly the same outside events and environmental circumstances could turn out to be widely different from one another and to take place at widely different times. This created a considerable problem in identifying general regularities on which to base inductive predictions. Had we wanted to predict, for instance, how the persecution of the Jews in Germany would develop, we would not have known where to find the relevant historical and sociological material, the 'series of cases' on which to base the explanatory and predictive covering generalisations for the hypotheses relating to the case in question. When in all human history had there ever been persecutions comparable to this one in nature and extent?[160]

Moreover, the behavioural models and actions performed by human groups were strongly conditioned by technological advance, which was itself dependent at root on the experimental and creative talents of the human mind. 'Prediction' would in these circumstances require the ability to anticipate something in itself highly improbable – scientific invention. To predict in advance Einstein's calculations on relativity, for example, would require 'anticipating' Einstein himself.[161] Even if such a thought were hypothetically thinkable, it would still be at variance with the historically well established fact that a close connection exists between the theoretical stances (and technologies) of a given period and the general shape of its social organisation (*Gesamtlebensform*).[162] Although scientific invention was unpredictable, it was all the same, at least in part, 'post-dictable': a given social context represented the actual sum of necessary, even if not sufficient, conditions for the arising of new scientific theories and new technologies. How could a society predict the invention of the wheel while it was still incapable of inventing it?[163] It was for these reasons too that long-term historical predictions proved to be so problematical.[164]

A second class of difficulties restricting the possibility of social prediction concerned the 'dispersion phenomena' (*Streuungserscheinungen*)[165] and 'instability phenomena' (*Labilitätserscheinungen*)[166] inherent in human behaviour. However much the social structure of a group conditioned the behaviour of individuals within it and tended to assimilate them to each other within fixed patterns of distribution and stratification of living standards (so aiding statistical prediction of average behaviour), the existence of individuals strongly deviating from the norm could not be overlooked. Not only was it in the nature of statistical generalisation that it did not extend to predictive extrapolation of the behaviour of individuals[167] but the existence above all of 'individual pecularities' made it extremely difficult to predict a person's career step by step.[168] One might be dealing, as in the case of Napoleon, with an individual capable of exerting exceptional influence over the course of history.

The fates of individuals were also very often affected by instability phenomena. It might happen that a chance alteration, however small, in the initial conditions of an individual's existence could lead to important consequences in the course of his life.[169] Instability phenomena could also affect whole groups and relate to a wide series of interconnected initial states, with the result that the behaviour of groups could not even be claimed in general to be more easily predictable than that of individuals.[170]

There was also, Neurath argued, a third specific limitation to sociological prediction. Here the theme of the paradoxes of sociological prediction, later to be widely developed by Robert Merton (followed by writers such as A. Grünbaum, E. Grünberg, F. Modigliani, P. Suppes, M. Scriven, and W.I. Thomas)[171] and to meet with a hostile reception from Popper in *The Poverty of Historicism*,[172] received a notable introduction. This was the limitation created by the reflexivity of social prediction: while the prediction of an eclipse did not influence the phenomenon itself, a prediction relating to market conditions could well influence the Stock Exchange, and a prediction of the imminence of revolution could well have the effect of helping or hindering its actual occurrence.[173]

A person who formed a prediction and then made it generally known (or conducted himself in such a manner as to make clear what his unspoken prediction was) could succeed reflexively in this way in strengthening his own prediction, so producing the 'self-fulfilling' prediction. Conversely the 'self-defeating' prediction was also possible: announcement of the prediction sets off a process of negative feedback in which the social agents affected by the predicted phenomenon react by carrying out behaviour contrary to the prediction.[174] In both of these cases the sociologist had to regard his own prediction as just one element in the initial conditions of the phenomenon he wished to study and predict: he had to work self-referentially by attempting an advance calculation of the complex play of interactions between expectations and expectations of expectations which the prediction was bound to engender.[175] All of which only added to the difficulty and complexity of his task.[176]

The 'unpredictability within empiricism',[177] Neurath ended by saying, was a bitter pill to swallow[178] for many enthusiasts of empirical sociology (especially economists) who would prefer to be able to see the difficulties of social prediction as only temporary and arising from the as yet not fully developed techniques of sociological investigation and our provisional 'incomplete knowledge of the situation'.[179] Neurath's reference here was clearly and importantly to the views argued in such works as *Foundations of Sociology* and *Social Research* by Georg A. Lundberg, the influential Swedish-American sociologist closely linked to neopositivistic thinking.[180] We were, Neurath argued, deceiving ourselves if we thought that the problems of sociological prediction arose from accidental difficulties, to be naturally overcome by more sophisticated techniques of inquiry. It ought to be openly admitted that theorists of empiricism had not grasped the deep-seated nature of the problems which affected the social sciences despite the attempt of Mill and his school to include them among the empirical disciplines. The trickiest problems arose from the fact that the more detailed aspects of human sociology involved taking language as an 'important social item' as well as accounting for the high degree of 'adaptability, flexibility, and alterability' which sharply differentiated the socio-cultural systems of *homo sapiens* from all other objects of empirical research.[181]

The scant attention which the empirical tradition had always given to these special facets of human behaviour had left the supporters of metaphysics with a splendid opportunity to foist their views on the human sciences.[182] Only social scientists who were capable of maintaining the same role in the historical and social sciences as thinkers such as Mach, Duhem, and Poincaré had in physics would, he concluded, be able to redeem their subject from the excesses of dogmatic empiricism and the mortgage of spiritualistic metaphysics.[183]

7. NEURATH AND HEMPEL

If the description of Neurath's sociological epistemology given in the preceding sections is correct, we have arrived at a considerably different interpretation of his thought from that advanced by Hempel.

In the first place it no longer seems possible that Neurath's opposition to the historicist division between *Naturwissenschaften* and *Geisteswissenschaften* manifested itself as a materialistic stance in his philosophical thought and thus as an unconditional adherence to behaviourism. Nor does it seem possible that this was the real reason for his criticism of Weber. This interpretation of Hempel's may receive some support from Neurath's early works, but it becomes far less convincing in the context of the works of his maturity, especially *Foundations of the Social Sciences.*

In these works we have found Neurath's rejection of the opposition between 'science of the spirit' and 'science of nature' and distrust of the mentalistic approach to have both an epistemological foundation and a pragmatic intention, and not to go back to a 'philosophical' espousal of materialism. What really concerned Neurath was the epistemological criticism of metaphysical ontologism and the refutation of a speculative or contemplative conception of the social sciences. The social sciences interested him for the means they offered for social intervention and planning. Because only an empirical sociology was in a position to fulfil this role in as much as it alone was able, under certain conditions, to make credible predictions, Neurath demanded that the language and procedures of social science should be as a matter of principle susceptible to empirical testing. His constant criticism of spiritualism and the philosophy of values therefore contained this pragmatic point: the use of mentalistic categories (which he saw as useful for comprehending the meaning of social action) could not be taken to lead to the use of intuition in sociology, as was claimed by the most radical supporters of *verstehende Soziologie*, since this removed it from all intersubjective testability. Only to this limited degree did Neurath share the behaviouristic position, of whose other aspects, as we have seen, he was openly critical, just as he criticised the excesses of empiricism and the tendency towards the overmathematisation of sociology.

Attempts to see Neurath as being close to Watson, Pavlov, or Gilbert Ryle are therefore misleading. His position, as expressed in *Foundations of the Social Sciences*, was much closer to that of Weber. In some ways it even appears more conventionalistic and relativistic than Weber's. For Neurath the epistemological choice in the social sciences did not lie between *Verstehen* and *Erklären*, since he denied that the sociologist's job lay in the causal explanation of social phenomena, as Weber had argued it did (although also attributing an idiographic function to causal explanation). For Neurath the sociologist's job was to utilise the processes of *Verstehen* to interpret the complexity of social phenomena and to formulate, within the bounds of local and temporal rationality, 'valid' predictions, provisional though they inevitably were and based on conjectural hypotheses. *Verstehen* and *Erklären* were not diametrically opposed to one another: rather idiographic understanding and nomological research were both directed towards prediction, *Voraussehen.* It was in this sense that Neurath expressed the hope that thinkers would arise in the social sciences capable of fulfilling the same function as Mach, Duhem, and Poincaré had fulfilled in the natural sciences.

Secondly the definition of the premises of value in Neurath's sociology appears more problematical than Hempel has succeeded in showing. That Marxism exercised an important and lasting influence on Neurath's thought cannot be doubted, but his brand of Marxism was not only not dogmatic, but essentially heterodox and anti-Leninist. Moreover, in addition to the Marxist Neurath, there is the pre-Marxist Neurath – the Neurath of social enlightenment and the 'scientific utopian' – and there is the post-Marxist Neurath – supporter of political pluralism, individual liberty and civil tolerance, critic of centralised technocratic planning, and deviser of a 'planning for freedom'. The premises of value were, however, openly stated in Neurath's work, as was not the case with Carnap – a difference of some epistemological importance. What Hempel somewhat ungraciously decries as a lack of analytical and systematic rigour or as 'ideologism' on Neurath's part was in fact the product of a clear epistemological awareness. Such systematic research as Carnap advocated on the precision of language and the construction of ideal languages in fact struck Neurath as sterile on the theoretical no less than the practical level.

Hempel's admiration for Carnap's rigour and quality of analysis and disdain of Neurath's 'ideological vagueness' shows quite unequivocally the prejudice behind his evaluation. Despite his valuable contribution to the liberalising of empiricism, Hempel showed himself at the end of the Sixties, when he wrote 'Logical Positivism and the Social Sciences', even more attached to the orthodoxy of Viennese logical positivism than Neurath had ever been at the start of the Forties. He appeared still to subscribe to Carnap's position on the purity of philosophical methods and the perfect ideological and moral neutrality of theoretical language.[184]

Thirdly, there can be no doubt that Neurath's sociology was typified by a strongly pragmatic and instrumentalistic orientation, that it aimed exclusively at prediction of social phenomena and not at their explanation, and this is justifiably emphasised by Hempel. But it would be fair also to emphasise the extreme caution with which Neurath argued for the predictive role of the social sciences and not to overlook the way in which he saw the many problems and paradoxes as constituing no less real limitations to the social than to the natural sciences. Nor was this because of any putative immaturity of the social sciences in comparison with the 'exact sciences', but because of the greater difficulty of their task and the extreme complexity of their fields of study. The thought of Neurath and Hempel can only be viewed as being greatly at variance here.

Neurath assumed the natural sciences also within a historical and sociological perspective and assigned pragmatic functions of prediction to the whole complex of the empirical disciplines, conscious both of the impossibility of a rigorous scientific 'explanation' and of the problems inherent in prediction. Hempel, on the other hand, attributes both to the natural and to the social sciences the task not only of predicting, or attempting to predict, phenomena, but also of explaining them. Herein lies the significance of the deductive-nomological model of scientific explanation which he formulated for physics and which he has since proposed to extend to the social sciences, including historiography.[185] Here too we see the unbridgeable gap which separates the 'liberalised' logical empiricism of the chief exponent of the standard view in epistemology in the Fifties from the 'sociological', conventionalistic, and pragmatic empiricism of Neurath.

NOTES

[1] See in particular C.G. Hempel, *Aspects of Scientific Explanation*, New York: The Free Press, 1970. For an accurate bibliography of Hempel's writings see N. Rescher (ed.), *Essays in Honor of Carl G. Hempel*, Dordrecht: D. Reidel, 1969, pp. 266–70.
[2] C.G. Hempel, 'Logical Positivism and the Social Sciences', pp. 165–74. See also C.G. Hempel, 'Schlick und Neurath: Fundierung vs. Kohärenz in der wissenschaftlichen Erkenntnis', pp. 10–4.
[3] C.G. Hempel, 'Logical Positivism and the Social Sciences', p. 166.
[4] Ibid., p. 169. Cf. G. Ryle, *The Concept of Mind*, London: Hutchinson's University Library, 1949.
[5] C.G. Hempel, 'Logical Positivism and the Social Sciences', pp. 168–9, 171–2.
[6] Ibid., p. 169; ("Neurath's writings on this subject often seem more like political manifestoes, like programs both for analysis and for action, than like carefully reasoned analytic studies", ibid., pp. 166–7).
[7] Ibid., p. 173.
[8] Cf. F. Suppe, *The Structure of Scientific Theories*, pp. 230–2, 619–24, 713–67.
[9] Cf. M. Scriven, 'Logical Positivism and the Behavioral Sciences', in P. Achinstein and S.F. Barker (eds.), *The Legacy of Logical Positivism*, p. 197.
[10] It should be noted that the sources for Hempel's reconstruction of Neurath's sociological thought amount only to the following: *Empirische Soziologie*, *Einheitswissenschaft und Psychologie*, 'Soziologie im Physikalismus', *Foundations of the Social Sciences*.
[11] O. Neurath, 'Zur Theorie der Sozialwissenschaften', [1910], *GpmS*, pp. 23–46.
[12] Ibid., p. 26. K. Pearson's book *The Grammar of Science* (London: Adam and

Charles Black, 1900) attracted attention from Mach in *Erkenntnis und Irrtum* (1906) and sharp polemic from Lenin in *Materialism and Empiriocriticism* (1909).

[13] O. Neurath, *Empirische Soziologie*, [1931e], *GpmS*, pp. 424–527, *ES*, pp. 319–421.

[14] O. Neurath, 'Soziologie im Physikalismus', [1931f], pp. 393–431, *PP*, pp. 58–90.

[15] O. Neurath, *Empirische Soziologie*, [1931e], *GpmS*, pp. 465–70, *ES*, pp. 358–63.

[16] Ibid., *GpmS*, p. 469, *ES*, p. 361.

[17] Ibid., *GpmS*, pp. 455–65, *ES*, pp. 353–8; O. Neurath, 'Soziologie im Physikalismus', [1931f], pp. 405–14, *PP*, pp. 68–75; O. Neurath, 'Physikalismus', [1931d], p. 302, *PP*, p. 56; O. Neurath, 'Bürgerlicher Marxismus', *Der Kampf*, *23* (1930), now in *GpmS*, p. 352; O. Neurath, 'Weltanschauung und Marxismus', *Der Kampf*, *24* (1931), now in *GpmS*, p. 409.

[18] O. Neurath, 'Soziologie im Physikalismus', [1931f], pp. 408, 409, 411, *PP*, pp. 70, 71, 73; O. Neurath, 'Sozialbehaviorismus', *Soziologus*, *8* (1932), now in *GpmS*, p. 567; O. Neurath, *Empirische Soziologie*, [1931e], *GpmS*, pp. 463–4, *ES*, p. 357.

[19] Ibid., *GpmS*, p. 463, *ES*, p. 357.

[20] O. Neurath, 'Soziologie im Physikalismus', [1931f], pp. 408–9; *PP*, pp. 71–2; O. Neurath, 'Mensch und Gesellschaft in der Wissenschaft', in *Actes du Congrès internationale de philosophie scientifique, Sorbonne, Paris 1935*, II, *L'unité de la science*, in *Actualités Scientifiques et Industrielles*, no. 389, Paris 1936, now also in *GpmS*, p. 714.

[21] Cf. F.R. Dallmayr and T.A. McCarthy, *Understanding and Social Inquiry*, Notre Dame (Ind.): University of Notre Dame Press, 1977, p. 6. See also: T.A. McCarthy, 'On Misunderstanding 'Understanding'', *Theory and Decision*, *3* (1973), pp. 351–70; T.A. McCarthy, 'The Operation Called Verstehen: Towards a Redefinition of the Problem', in *Boston Studies in the Philosophy of Science*, vol. XX, ed. by K.S. Schaffner and R.S. Cohen, 1974, pp. 167–93; T. Abel, 'The Operation Called Verstehen', *American Journal of Sociology*, *54* (1958), pp. 211–8; A.W. Diquattro, 'Verstehen as an Empirical Concept', *Sociology and Empirical Research*, *57* (1972), p. 35.

[22] Cf. G.H. von Wright, *Explanation and Understanding*, Ithaca and New York: Cornell University Press, pp. 31, 182, note 91.

[23] O. Neurath, 'Soziologie im Physikalismus', [1931f], p. 408, *PP*, pp. 70–1; O. Neurath, *Empirische Soziologie*, [1931e], *GpmS*, p. 455, *ES*, pp. 353–4; O. Neurath, 'Soziologische Prognosen', [1936h], p. 400; O. Neurath, 'Bürgerlicher Marxismus', [1930], *GpmS*, p. 352; O. Neurath, 'Weltanschauung und Marxismus', [1931a], *GpmS*, p. 410; O. Neurath, Review of W. Sombart, *Die drei Nationalökonomien*, in *Der Kampf*, *23* (1930), now in *GpmS*, p. 357; O. Neurath, *Einheitswissenschaft und Psychologie*, [1933], *GpmS*, p. 603.

[24] O. Neurath, 'Soziologie im Physikalismus', [1931f], pp. 405–8, *PP*, pp. 68–71; O. Neurath, *Einheitswissenschaft und Psychologie*, [1933], *GpmS*, p. 603.

[25] O. Neurath, *Empirische Soziologie*, [1931e], *GpmS*, pp. 461–3, *ES*, pp. 356–8.

[26] O. Neurath, 'Soziologie im Physikalismus', [1931f], p. 409, *PP*, pp. 71–2.

[27] O. Neurath, *Empirische Soziologie*, [1931e], *GpmS*, p. 431, *ES*, pp. 325; O. Neurath, 'Soziologie im Physikalismus', [1931f], pp. 408–11, *PP*, pp. 71–3.

[28] Ibid., pp. 408–10, *PP*, pp. 71–3; O. Neurath, *Empirische Soziologie*, [1931e], *GpmS*, pp. 454, 470, *ES*, p. 325.

[29] O. Neurath, 'Soziologie im Physikalismus', [1931f], pp. 409–11, *PP*, pp. 71–3.

[30] Ibid., pp. 409, 410, *PP*, pp. 71, 72; O. Neurath, *Empirische Soziologie*, [1931e], *GpmS*, pp. 468, 482, *ES*, pp. 361, 374.
[31] O. Neurath, 'Soziologie im Physikalismus', [1931f], p. 412, *PP*, p. 73; O. Neurath, 'Sozialbehaviorismus', [1932], *GpmS*, p. 567.
[32] O. Neurath, 'Soziologie im Physikalismus', [1931f], p. 412, *PP*, p. 73.
[33] Ibid., loc. cit.; O. Neurath, *Empirische Soziologie*, [1931e], *GpmS*, p. 468, *ES*, pp. 361.
[34] O. Neurath, 'Soziologie im Physikalismus', [1931f], p. 411, *PP*, p. 73.
[35] O. Neurath, 'Mensch und Gesellschaft in der Wissenschaft', [1936b], *GpmS*, pp. 714–5.
[36] Ibid., loc. cit.; O. Neurath, 'Soziologie im Physikalismus', [1931f], pp. 426, 428, *PP*, pp. 85–6, 87.
[37] Cf. M. Weber, *Wirtschaft und Gesellschaft*, English trans. pp. 104–5, 106–7. Weber considers Neurath's analysis of problems concerning the 'natural calculus' of economic values to be 'particularly penetrating'. On the other hand he expresses his deep disagreement about 'socialisation': cf. Weber's letter to Neurath of 4th December 1919, now in E. Baumgarten, *Max Weber: Werk und Person*, Tübingen: Mohr, 1964.
[38] O. Neurath, *Empirische Soziologie*, [1931e], *GpmS*, p. 461, *ES*, pp. 356–7.
[39] Ibid., *GpmS*, p. 464, *ES*, p. 357.
[40] Ibid., *GpmS*, p. 463, *ES*, p. 357.
[41] Ibid., *GpmS*, p. 464, *ES*, p. 357.
[42] O. Neurath, 'Soziologie im Physikalismus', [1931f], p. 423, *PP*, p. 83.
[43] O. Neurath, *Empirische Soziologie*, [1931e], *GpmS*, p. 464, *ES*, p. 358.
[44] Cf. A. Fanfani, *Catholicism, Protestantism and Capitalism*, English trans. by T. Parsons, London: Sheed & Ward, 1935. Cf. Neurath's quotation of this book in *Foundations*, [1944a], p. 48.
[45] Cf. J.B. Kraus, *Scholastik, Puritanismus und Kapitalismus*, München: Duncker & Humblot, 1930. See also Neurath's review of this book under the title 'Marxismus eines Jesuiten', in *Der Kampf*, *24* (1931), pp. 271–4, now in *GpmS*, pp. 401–5.
[46] O. Neurath, 'Soziologie im Physikalismus', [1931f], pp. 423–4, *PP*, pp. 83–4.
[47] O. Neurath, *Empirische Soziologie*, [1931e], *GpmS*, pp. 464, 494, *ES*, pp. 358, 384.
[48] Ibid., *GpmS*, p. 464, *ES*, pp. 358.
[49] O. Neurath, 'Soziologie im Physikalismus', [1931f], p. 428, *PP*, p. 87. Neurath considered E. Ciccotti's book, *Il tramonto della schiavitù nel mondo antico*, Torino: Bocca, 1899 (German trans. *Der Untergang der Sklaverei im Altertum*, ed. by O. Oldberg, Berlin: Vorwärts, 1910) as an example of Marxist sociological research (ibid.).
[50] Josef Popper-Lynkeus (1838–1931), who was a close friend of Mach, had written several political-philosophical essays, among them *Phantasien eines Realisten* (1900), *Voltaire* (1905), *Die allgemeine Nährpflicht als Lösung der sozialen Frage* (1912); see O. Neurath, 'Josef Popper-Lynkeus, seine Bedeutung als Zeitgenosse', *Neues Frauenleben*, *20* (1918), no. 3, pp. 33–8, now in *GpmS*, pp. 131–6.
[51] Carl Ballod-Atlanticus (1964–1933) was the author of *Grundriße der Statistik* (1913) and *Die Zukunftstaat* (1919).
[52] Neurath knew Walter Rathenau's most important works, particularly the famous essays *Zur Kritik der Zeit* (1912) and *Zur Mechanik des Geistes* (1913).

[53] O. Neurath, *Durch die Kriegswirtschaft zur Naturalwirtschaft*, München: Georg D.W. Callwey, 1919, English trans. *ES*, pp. 133, 140, 151, 152–3; O. Neurath, *Lebensgestaltung und Klassenkampf*, [1928], *GpmS*, pp. 237–43, *ES*, pp. 259–66; O. Neurath, *Empirische Soziologie*, [1931e], *GpmS*, p. 443, *ES*, p. 339; O. Neurath, *Le développement du Cercle de Vienne et l'avenir de l'empirisme logique*, [1936], p. 41; O. Neurath, 'Utopien', *Wirtschaft und Lebensordnung, 1* (1919), pp. 5–6, now in *GpmS*, pp. 137–8. See in addition: L. von Mises, 'Die Wirtschaftrechnung im sozialistischen Gemeinwesen', *Archiv für Sozialwissenschaft und Sozialpolitik, 48* (1920–21), pp. 102–9; F.A. Hayek, 'Scientism and the Study of Social Sciences', in F.A. Hayek, *The Counter-Revolution of Science*, Glencoe (Ill.): The Free Press, 1955, pp. 45, 96, 219; F.A. Hayek, *Collectivist Economic Planning*, London: G. Routledge, 1935, pp. 30–1. See in addition: C. Ruini, *L'economia di guerra. Linee teoriche*, Bari: Laterza, 1940, pp. 11–8.

[54] For a selection of Neurath's economic texts and a somewhat different interpretation of his use of Marxism, with reference both to his earliest economic works (pre–1914) and also those of the post-war period, see O. Neurath, *Economic Writings*, ed. by R.S. Cohen and M. Neurath, Dordrecht and Boston: D. Reidel, 1989 (forthcoming). See also: R.S. Cohen, 'Neurath, Otto', in *The Encyclopedia of Philosophy*, ed. by P. Edwards, New York: Macmillan, 1967; R. Hegselmann, 'Otto Neurath. Empiristischer Aufklärer und Sozialreformer', pp. 19–23, 24–33; E. Nemeth, *Otto Neurath und der Wiener Kreis*, pp. 11–50; K. Fleck, *Otto Neurath. Eine biographische und systematische Untersuchung*, pp. 14–23, 30–68; E. Mohn, *Der logische Positivismus. Theorien und politische Praxis seiner Vertreter*; R. Gerlich, *Die gescheiterte Alternative: Sozialisierung in Österreich nach dem Ersten Weltkrieg*, Wien: Braumüller, 1980, pp. 42 ff; F. Weber, 'Otto Neurath und die Sozialisierung', in F. Stadler (ed.), *Arbeiterbildung in der Zwischenkriegszeit*, pp. 133–9; E. Wissel, 'L'Internazionale socialista e il dibattito sulla socializzazione', in *Storia del marxismo*, III, Torino: Einaudi, 1980, p. 208.

[55] On the relationships between the Vienna Circle and Austro-Marxism see E. Nemeth, *Otto Neurath und der Wiener Kreis*, pp. 55–60. See in addition: T. Bottomore's 'Introduction' to T. Bottomore and P. Good (eds.), *Austro-Marxism*, Oxford: Clarendon Press, 1978, pp. 1–44; F. Stadler, *Vom Positivismus zur 'Wissenschaftlichen Weltauffassung'*, pp. 81 ff; E. Glaser, 'Otto Neurath und der Austromarxismus', in F. Stadler (ed.), *Arbeiterbildung in der Zwischenkriegszeit*, pp. 126–32; M.W. Wartofsky, 'Positivism and Politics. The Vienna Circle as a Social Movement', in R. Haller (ed.), *Schlick und Neurath. Ein Symposion*, pp. 79–101.

[56] Neurath explicitly criticised dialectical materialism: see for instance: O. Neurath, *Empirische Soziologie*, [1931e], *GpmS*, p. 524, *ES*, p. 415; O. Neurath, *Foundations*, [1944a], p. 48, note 28; and cf. R. Carnap, 'Intellectual Autobiography', in P.A. Schilpp (ed.), *The Philosophy of Rudolf Carnap*, p. 24. For a criticism of Neurath's sociology from a dialectical and historicist standpoint see M. Horkheimer, 'Die neueste Angriff auf die Metaphysik', *Zeitschrift für Sozialforschung, 6* (1937), English trans. in M. Horkheimer, *Critical Theory*, New York: Herder and Herder, 1972, pp. 132–87. For a sympathetic discussion of theoretical tensions affecting dialectical materialism and logical positivism see R.S. Cohen, 'Dialectical Materialism and Carnap's Logical Empiricism', in P.A. Schilpp (ed.), *The Philosophy of Rudolf Carnap*, pp. 99–158.

[57] See for instance: M. Adler, *Kausalität und Teleologie im Streite um die Wissenschaft*, Wien: Wiener Volksbuchh., 1904, where for the first time Marxism was developed as an empirical social science; M. Adler, 'Mach und Marx', *Archiv für Sozialwissenschaft und Sozialpolitik, 33* (1911), pp. 398–400; M. Adler, *Der soziologische Sinn der Lehre von Karl Marx*, Leipzig: C.L. Hirschfeld, 1914; M. Adler, *Kant und der Marxismus*, Aalen: Scientia Verlag, 1925. Adler criticised the sociological intuitionism and the epistemological theory of *Verstehen*. Austro-Marxism derived its epistemological concerns from Viktor Adler and Carl Grünberg. Also Friedrich Adler wrote an important essay on Mach: *Ernst Machs Überwindung des mechanischen Materialismus*, Wien: Wiener Volksbuchh., 1918. It is rather difficult to reconstruct Neurath's connections with the radical wing of mid-European Social Democracy, for instance with Karl Korsch and Anton Pannekoek, who were, like Neurath, deeply concerned with the problem of 'socialisation', with logical empiricism and the refutation of Lenin's dialectical materialism. See on this subject Korsch and Pannekoek's essays which appeared in *Erkenntnis, 3* (1932–33), pp. 389–400, and *8* (1939–40), pp. 397–402.

[58] Cf. G. Statera, *Logica, linguaggio e sociologia. Studio su Otto Neurath e il neopositivismo*, p. 137.

[59] O. Neurath, *Empirische Soziologie*, [1931e], *GpmS*, pp. 448–55, *ES*, p. 345–53; O. Neurath, 'Soziologie im Physikalismus', [1931f], pp. 421–8, *PP*, pp. 82–7; O. Neurath, 'Weltanschauung und Marxismus', [1931a], *GpmS*, p. 411.

[60] O. Neurath, *Empirische Soziologie*, [1931e], *GpmS*, pp. 449, 454–5, *ES*, pp. 346, 352–3.

[61] O. Neurath, 'Soziologie im Physikalismus', [1931f], pp. 418–9, *PP*, pp. 79; O. Neurath, 'Weltanschauung und Marxismus', [1931a], *GpmS*, p. 411.

[62] O. Neurath, 'Soziologie im Physikalismus', [1931f], p. 419, *PP*, p. 79.

[63] Ibid., loc. cit.

[64] Ibid., pp. 426–7, *PP*, p. 86.

[65] Ibid., p. 424, *PP*, p. 84.

[66] Ibid., pp. 426, 427, *PP*, pp. 85–6, 87.

[67] Ibid., p. 419, *PP*, pp. 79–80.

[68] Ibid., loc. cit., *PP*, p. 80; O. Neurath, *Empirische Soziologie*, [1931e], *GpmS*, pp. 469–70, *ES*, pp. 352; O. Neurath, *Einheitswissenschaft und Psychologie*, [1933], *GpmS*, p. 606.

[69] O. Neurath, 'Soziologie im Physikalismus', [1931f], p. 414, *PP*, pp. 75–6.

[70] Ibid., pp. 414–5, *PP*, p. 76.

[71] Ibid., p. 415, *PP*, p. 76; O. Neurath, 'Sozialbehaviorismus', [1932], *GpmS*, p. 567.

[72] O. Neurath, 'Soziologie im Physikalismus', [1931f], p. 423, *PP*, p. 83.

[73] Ibid., pp. 414, 425, *PP*, pp. 75, 85.

[74] Ibid., p. 425, *PP*, p. 85.

[75] O. Neurath, *Empirische Soziologie*, [1931e], *GpmS*, p. 478, *ES*, p. 371.

[76] Ibid., *GpmS*, pp. 448, 449–51, *ES*, pp. 345, 346–9.

[77] O. Neurath, *Lebensgestaltung und Klassenkampf*, [1928], *GpmS*, p. 289, *ES*, p. 293.

[78] Ibid., loc. cit.

[79] Ibid., loc. cit.

[80] Ibid., loc. cit.

[81] Ibid., loc. cit.
[82] O. Neurath, 'Soziologie im Physikalismus', [1931f], p. 429, *PP*, p. 88.
[83] Ibid., loc. cit.
[84] O. Neurath, *Empirische Soziologie*, [1931e], *GpmS*, p. 514, *ES*, p. 403.
[85] Ibid., *GpmS*, pp. 516–7, *ES*, pp. 406–7.
[86] O. Neurath, 'Soziologie im Physikalismus', [1931f], pp. 429–30, *PP*, pp. 88–9.
[87] O. Neurath, 'Bürgerlicher Marxismus', [1930], *GpmS*, pp. 349–50.
[88] Ibid., pp. 351, 354; O. Neurath, 'Weltanschauung und Marxismus', [1931a], GpmS, pp. 407–8 ff.
[89] G. Statera (*Logica, linguaggio e sociologia. Studio su Otto Neurath e il neopositivismo*, pp. 160 ff) and A. Pasquinelli (*La sociologia del positivismo*, pp. 41–8) point out the theoretical originality and importance of this essay with respect to Neurath's sociological works of the Thirties.
[90] O. Neurath, 'Universal Jargon and Terminology', [1941], pp. 142–3, *PP*, p. 225.
[91] O. Neurath, *Foundations*, [1944a], pp. 4–5.
[92] Ibid., pp. 18, 6; O. Neurath, 'Die neue Enzyklopädie', [1938c], *GpmS*, p. 870.
[93] O. Neurath, *Foundations*, [1944a], p. 6.
[94] O. Neurath, *Foundations*, [1944a], p. 18.
[95] Ibid., pp. 15, 14.
[96] O. Neurath, 'Physikalismus und Erkenntnisforschung', I, [1936f], *GpmS*, p. 755, *PP*, p. 164.
[97] O. Neurath, *Foundations*, [1944a], p. 16.
[98] Ibid., loc. cit.; O. Neurath, 'Wissen und Sein', [1938a], *GpmS*, p. 838.
[99] O. Neurath, *Foundations*, [1944a], p. 17.
[100] Ibid., pp. 13, 48, note 19. ("I think that 'semantics', as evolved by Carnap and Tarski, will support many kinds of calculus analysis, but I feel uneasy when thinking of its application to empiricist arguments and the danger of slipping into 'ontological' ways of arguing").
[101] Ibid., p. 12.
[102] Ibid., pp. 12, 11.
[103] Ibid., pp. 12, 47–8, note 16. Cf. J. Dewey and A.F. Bentley, *Knowing and the Known*, Boston: The Beacon Press, 1960, pp. 40, note 3, 225, note 7.
[104] O. Neurath, *Foundations*, [1944a], p. 12.
[105] Ibid., loc. cit.
[106] Ibid., p. 13.
[107] Ibid., loc. cit.
[108] O. Neurath, 'Prediction and Induction', [1946a], p. 3, *PP*, p. 244; O. Neurath, 'Wissen und Sein', [1938a], *GpmS*, pp. 839–43.
[109] O. Neurath, *Foundations*, [1944a], p. 13.
[110] Ibid., pp. 13, 14.
[111] Ibid., p. 14.
[112] Ibid., loc. cit.; O. Neurath, 'Universal Jargon and Terminology', [1941], pp. 129, 134, *PP*, pp. 215, 219.
[113] O. Neurath, *Foundations*, [1944a], p. 14.
[114] Ibid., pp. 35, 19, 34, 33.
[115] Ibid., pp. 34, 33.
[116] O. Neurath, *Empirische Soziologie*, [1931e], *GpmS*, pp. 509–10, *ES*, pp. 398–9; O.

Neurath, 'Zur Theorie der Sozialwissenschaften', [1910], *GpmS*, pp. 35–6.
117 O. Neurath, *Foundations*, [1944a], pp. 35, 34, 24, 33.
118 Ibid., pp. 32–3.
119 Ibid., p. 33.
120 Ibid., pp. 32–5.
121 Ibid., pp. 12–3, 24.
122 Ibid., p. 25.
123 Ibid., p. 24.
124 Ibid., p. 25; cf. P. Duhem, *La théorie physique: son object, sa structure*, pp. 278–89, English trans. pp. 183–90.
125 O. Neurath, *Foundations*, [1944a], p. 25.
126 Ibid., p. 26.
127 Ibid., p. 25.
128 In criticising causalism Neurath appealed to the authority of P. Frank (*Das Kausalgesetz und seine Grenzen*, Wien: Julius Springer, 1931) and W.H. George (*The Scientist in Action: A Scientific Study of his Method*, London: Williams & Norgate, 1936).
129 O. Neurath, *Foundations*, [1944a], p. 22.
130 O. Neurath, *Foundations*, [1944a], p. 20. Neurath quoted J.W. Bews, *Human Ecology*, London: Oxford University Press, 1935, and C. Elton, *Animal Ecology*, London: Sidgwick & Jackson, 1927.
131 O. Neurath, *Foundations*, [1944a], p. 22.
132 Ibid., p. 20.
133 Ibid., loc. cit.; O. Neurath, 'Universal Jargon and Terminology', [1941], pp. 134–7, *PP*, pp. 218–20.
134 O. Neurath, 'Universal Jargon and Terminology', [1941], p. 134, *PP*, p. 218.
135 O. Neurath, 'The Orchestration of the Sciences by the Encyclopedism of Logical Empiricism', [1945/46], p. 498, *PP*, p. 232.
136 O. Neurath, *Foundations*, [1944a], pp. 20–1.
137 Ibid., p. 30.
138 Ibid., loc. cit.
139 Ibid., loc. cit.
140 Cf.: P. Gardiner, *The Nature of Historical Explanation*, Oxford: Oxford University Press, 1961, pp. 8–9; W. Dray, *Laws and Explanation in History*, Oxford: Oxford University Press, 1957, pp. 110–2. On this subject even a severe critic of the empiricist epistemology such as William Dray maintains a more moderate thesis than Neurath.
141 O. Neurath, *Foundations*, [1944a], p. 21.
142 Ibid., loc. cit.; O. Neurath, 'Sozialbehaviorismus', [1932], *GpmS*, p. 568; O. Neurath, 'Weltanschauung und Marxismus', [1931a], *GpmS*, p. 411.
143 O. Neurath, 'Mensch und Gesellschaft in der Wissenschaft', [1936b], *GpmS*, p. 715.
144 Ibid., p. 714; O. Neurath, *Foundations*, [1944a], p. 22.
145 Ibid., loc. cit.
146 O. Neurath, 'Universal Jargon and Terminology', [1941], pp. 141–2, *PP*, pp. 224–5.
147 O. Neurath, *Empirische Soziologie*, [1931e], *GpmS*, pp. 437, 438, 499, *ES*, pp.

332, 333, 388.
[148] Ibid., pp. 500–14, *ES*, pp. 309–403.
[149] Ibid., pp. 438–9, *ES*, pp. 333–4.
[150] Ibid., pp. 439–42, *ES*, pp. 334–8.
[151] Ibid., pp. 511–4, *ES*, pp. 400–3; O. Neurath, 'Inventory of the Standard of Living', *Zeitschrift für Sozialforschung*, *6* (1937), pp. 140–51.
[152] Cf. K.R. Popper, *The Poverty of Historicism*, pp. 58–70.
[153] O. Neurath, *Empirische Soziologie*, [1931e], *GpmS*, pp. 525, 514, *ES*, pp. 417, 403.
[154] Ibid., *GpmS*, pp. 514 ff, *ES*, pp. 404 ff; O. Neurath, *Foundations*, [1944a], pp. 26–30.
[155] Ibid., p. 22.
[156] Ibid., p. 27.
[157] Ibid., p. 28.
[158] Ibid., p. 38.
[159] O. Neurath, *Empirische Soziologie*, [1931e], *GpmS*, pp. 515, 514, ("Wenn wir voraussagen wollen, was Völker, Staaten, Organisationen tun werden, können wir nicht auf die letzen physikalischen Elemente zurückgehen, wir müssen uns damit begnügen, gewisse grobe Tatbestände komplexer Art ins Auge zu fassen"), *ES*, p. 404.
[160] O. Neurath, *Foundations*, [1944a], p. 26. D. Thomas criticises Neurath's argument in *Naturalism and the Social Sciences*, Cambridge: Cambridge University Press, 1979, p. 18.
[161] O. Neurath, *Empirische Soziologie*, [1931e], *GpmS*, p. 515, *ES*, p. 405. In the Fifties identical arguments were emphatically presented by Popper as a 'strictly logical' refutation of historicism: see K.R. Popper, *The Poverty of Historicism*, pp. 1–14.
[162] O. Neurath, *Empirische Soziologie*, [1931e], *GpmS*, p. 515, *ES*, p. 405.
[163] O. Neurath, *Foundations*, [1944a], p. 30; O. Neurath, 'Soziologische Prognosen', [1936h], pp. 403–4.
[164] O. Neurath, *Empirische Soziologie*, [1931e], *GpmS*, pp. 515–6, *ES*, p. 405.
[165] Ibid., *GpmS*, p. 518, *ES*, p. 409.
[166] Ibid., *GpmS*, loc. cit., *ES*, p. 408; O. Neurath, *Foundations*, [1944a], p. 28.
[167] Ibid., p. 26.
[168] O. Neurath, *Empirische Soziologie*, [1931e], *GpmS*, p. 518, *ES*, p. 408.
[169] Ibid., loc. cit.
[170] O. Neurath, *Foundations*, [1944a], pp. 26–7.
[171] Cf. M. Brodbeck (ed.), *Readings in the Philosophy of the Social Sciences*, New York: The Macmillan Company, 1968, p. 756.
[172] Cf. K.R. Popper, *The Poverty of Historicism*, p. 13.
[173] O. Neurath, *Empirische Soziologie*, [1931e], *GpmS*, p. 516, *ES*, p. 405.
[174] Ibid., *GpmS*, loc. cit., *ES*, pp. 405–6; O. Neurath, *Foundations*, [1944a], pp. 28–9, 43.
[175] Ibid., pp. 28–9.
[176] Ibid., pp. 29, 43; O. Neurath, *Empirische Soziologie*, [1931e], *GpmS*, p. 516, *ES*, pp. 405–6; O. Neurath, 'Soziologische Prognosen', [1936h], p. 404.
[177] This is the title of one of the paragraphs of Neurath's *Foundations*, [1944a], pp. 28–30.

[178] O. Neurath, 'The Orchestration of the Sciences by the Encyclopedism of Logical Empiricism', [1945/46], p. 498, *PP*, p. 232.
[179] O. Neurath, *Foundations*, [1944a], p. 37; O. Neurath, 'The Orchestration of the Sciences by the Encyclopedism of Logical Empiricism', [1945/46], p. 498, *PP*, p. 232.
[180] Cf.: G.A. Lundberg, *Foundations of Sociology*, New York: The Macmillan Company, 1939, pp. 20–1; G.A. Lundberg, *Social Research*, New York: Green & Co., 1942, p. 21. See also G. Statera, 'Introduzione' to O. Neurath, *Sociologia e neopositivismo*, pp. 14–5.
[181] O. Neurath, *Foundations*, [1944a], pp. 37–8, 27. Popper devotes paragraph 25 of *The Poverty of Historicism* to a critique of Neurath's thesis that the variability of social conditions presents a specific difficulty for the social sciences (pp. 93–9).
[182] O. Neurath, *Foundations*, [1944a], p. 38.
[183] Ibid., loc. cit.; L. Kolakowski, in *Filozofia pozytywistyczna*, Warshau 1966, English trans. *Positivist Philosophy*, Harmondsworth: Penguin Books, 1972, pp. 221–2, 223, interprets Neurath's epistemology of the social sciences as a model of the most rigid positivism.
[184] Cf. C.G. Hempel, 'Science and Human Values', in C.G. Hempel, *Aspects of Scientific Explanation*, pp. 90–1.
[185] Cf.: C.G. Hempel, 'The Function of General Laws in History', in C.G. Hempel, *Aspects of Scientific Explanation*, pp. 231–41; C.G. Hempel, 'The Logic of Functional Analysis', ibid., pp. 297–330.

CHAPTER 7

EVALUATION, PRESCRIPTION, AND POLITICAL DECISION

1. TOWARDS A SOCIOLOGY OF SOCIOLOGY

The paradox of the self-fulfilling (or self-defeating) prediction was, Neurath pointed out, only one element within a still wider context – that of the circular inclusion of the sociologist's work as a part itself of the subject matter of his research. Epistemological examination of the social sciences, and especially of sociology, had to take its starting point within the self-referential circle in which the social scientist's activity was itself to be analysed as 'social praxis' and in which "the scientist himself figures as one of the elements of the social picture".[1]

The sociologist, and indeed any other scientist, did not enjoy a kind of 'social extraterritoriality' (*eine Art sozialer Exterritorialität*): as a scientist he "reasons, makes decisions, and functions like any other human being",[2] his behaviour being conditioned by the hopes and fears[3] inherent in his psychology as well as by his social position. His scientific work – in the form of the books he wrote, lectures he gave, and experiments he carried out – needed to be studied not only in terms of 'the psychology of knowledge' (*Behavioristik des Wissens*), but also in the context of the scientific community to which he belonged and the overall cultural mores of his time (*Gewohnheiten*).[4] Epistemology had therefore to become a 'sociology of knowledge' (*Soziologie des Wissens*).[5] And then, in the specific context of the social sciences, as well as being a preliminary and general condition of the sociology of knowledge, epistemology had to be understood and practised as a 'sociology of sociology'.[6] At the centre of this particular line of epistemological inquiry should stand "the language used by the sociologists themselves, with their statements and their habits".[7] A sociology of the sciences could not really be developed until a 'classification of the sociological hypotheses'[8] had first been sufficiently established on the basis of a 'behavioristic of sociologists'. And this was an area of research as yet in its infancy:

> we have fine questionnaires as far as preliterate tribes are concerned but hardly any when we try to ask sociologists how they themselves behave in arguing and writing.[9]

It was, moreover, a particularly delicate area: analysis of the relation between verbal behaviour and social context proved difficult enough under normal conditions, but was all the more so when the verbal behaviour of sociologists was concerned, and especially so in the case of sociological predictions on account of the reflexivity they involved.[10] The reason for this lay in the extent to which social scientists were conditioned more than any other scientists by 'taboos and old folklore' operating in a hidden manner which could not readily be gathered from the public documents of scientific communities.[11]

A sociological inquiry into the communities of social scientists would not merely reveal that 'value judgments' lay at the heart of social research: it would also show that no science could proceed without the preliminary uncritical acceptance of certain aspects of its own cultural tradition (or 'folklore').[12] To this extent, all scientists necessarily (and in some ways dogmatically) introduced elements of the taboos and traditions of their time into their specific areas of research. Weber's principle of *Wertfreiheit* could be understood therefore as a pragmatic maxim recommending tolerance and an inclination towards self-criticism as the distinguishing features of a modern scientific attitude, and as requiring the social scientist to overcome whatever elements of ideological factiousness he had.[13] It could not be understood, however, as a principle of absolute evaluative 'neutrality' on the part of the scientific worker.[14]

The evaluative assumptions and personal preferences of social scientists should therefore be recognised for what they were rather than be concealed or, worse, presented in the guise of 'eternal values'.[15] In this way, to the extent that they took account of their own relativity and subjectivity, the 'ideological' choices of scientists could, Neurath believed, play a positive role in scientific work. Certainly it was not unreasonable for 'wishful thinking' (that is, thought based on ethical and ideological prejudices which substituted wishes for reality and gave utopian expression to general requirements lacking empirical premises) to be confronted with the critical attitude encouraged by Weber. But it was necessary not to forget that, in the selection and in the very formulation of problems, "love and hate are often good teachers".[16] The intellectual enthusiasm of the researcher could teach him to see things others did not and suggest to him lines of approach which, although initially conditioned by his own interests and personal preferences, need not for this reason prevent him from employing properly scientific procedures.[17]

The social researcher's aim should not therefore be one of an ascetic lack of evaluation, but rather one of 'thinkful wishing'.[18] What Neurath had in mind with this intriguing coinage was an attitude which was both evaluative and critical at the same time, one which would lead the social scientist not to the faint-hearted rejection of his deepest wishes but to the systematic submission of them to the test of experience and to their sincere, open, non-propagandistic, and 'democratic' comparison with the theoretical theses and ideologies of his opponents.[19]

He recognised further that the passing-on of 'values' and traditions was a socially important function and was, *rebus sic stantibus*, one which could not be abandoned. But dressing one's individual subjective choices up in scientific clothes and smuggling them in as objective truths differed greatly from the ascribing of one's actions to moral and ideological (including religious) choices and commending them to succeeding generations, as happened in the course of education.[20] Accurate separation of theoretical propositions belonging to science from prescriptive propositions (or 'commandments') extraneous to the language of the empirical sciences was also quite different from dogmatically setting out to combat or ignore, as if purely emotional and irrational, everything evaluative or, like taboos and cultural folklore, belonging to the common inheritance of a people.[21]

2. SOCIAL THEORY, ETHICS, AND LAW: THEORETICAL PROPOSITIONS AND PRESCRIPTIVE PROPOSITIONS

Awareness that the scope for rigorous knowledge was extremely limited, together with the particular realisation that prediction presented serious difficulties in the social sciences did not constitute a reason, in Neurath's opinion, for the sociologist to risk emasculating his work by reducing the spread of his interests and occupying himself solely with 'social fossils', as sociologists and anthropologists did have a tendency to do.[22] The organisation and development of human life necessitated a constant degree of decision-making: decisions had to be made, and made expeditiously, on the basis of available scientific knowledge. Modern society constantly required the sociologist, as an expert in social problems, to put his knowledge at the service of the processes of collective decision-making, whether in economics, politics, law, or education. He could not take shelter, as other researchers did, in the speciality of his subject, pleading scientific neutrality and leaving the task of decision to others without giving them any scientific assistance. The decisions would be made all the same, most probably on the basis of false assumptions, and dogmatically or with irrationality.[23] The services of the sociologist, as social programmer and 'engineer', were indeed capable of raising the level of rationality in social decisions and he should not therefore remove himself from the performance of this social function.

But what kind of service could sociology provide for normative disciplines such as law, political economics, ethics, and education? Could sociology turn itself into a general 'normative science'? Was the social scientist in a position to give a normative and univocal indication of which decisions should be made in any given situation? Could he decide for others?

Neurath followed Hume closely in saying that the theoretical propositions of empirical sociology could not in general be taken as premises of normative propositions of an ethical, legal, or political kind. On the contrary, the task of the social scientist was precisely that of showing the lack of foundation in any empirical science which claimed to work with prescriptive propositions or through the prescriptive use of statements – "no science can teach what 'should' be done".[24]

In an early work – 'Zur Theorie der Sozialwissenschaften' of 1910 – Neurath had previously made a clear distinction between the ineradicable evaluative component present in any empirical proposition, in whatever area of science it belonged, and the 'suggestive function' (*Suggestion*) which certain statements could exercise independent of their syntactical structure and even of their propositional value.[25] In opposition to Weber, Neurath had included in the scope of scientific language such evaluative propositions or 'value judgments' (*Werturteile*) as "This person appeals to me" or "This thing is (i.e. appears to me to be) red". He had on the other hand judged as linguistically non-analysable the particular pragmatic function played by language when it was used with the intention of influencing (*beeinflußen*) others' behaviour.[26] In particular, this included statements constructed with an explicitly imperative syntax (*Imperative*), such as "Now you must do this". The suggestive/imperative function of (prescriptive) language could be studied in linguistic psychology and sociology, but not in the epistemological area of the formation of concepts and scientific theories. Words could

> be used as hisses, caresses, or blows, but when used in this way they can neither be consistent with theoretical propositions nor contradict them.[27]

Neurath's epistemology of the social sciences displayed therefore a clear and in many ways original distinction between 'value judgments' and 'prescriptive propositions', between 'axiological' and 'deontological'. The theoretical constructs of the researcher were necessarily conditioned by assumptions of value, in as much as these expressed unavoidable preferences on his part and reflected his social position and cultural habits. Imperatives, on the other hand, in as much as they generalised certain presuppositions of value and prescribed them as rules for individual or group behaviour, were outside the scope of the empirical sciences and theoretical language. The presence of premises of value, of linguistic conventions, of methodological decisions, or of pragmatic rules etc., in the procedures leading to formation of concepts in the social sciences, as in any other empirical science, attributed to scientific sentences – even when they formally presented a syntactical structure of a descriptive or assertive kind – the epistemological status of hypothetical/probabilistic propositions: "if we assume that A, then, if B, it is probable that C".[28] Sentences of empirical theories were never therefore 'factual propositions' capable of being verified or falsified, nor were they apophantic statements logically testable through truth functions. For the same reason they could not constitute the premise – unless purely pragmatic and informative – for the construction of prescriptive propositions:

> an imperative is never deducible from a system of theoretical propositions: this is not a limitation to the scientific method, but is simply the result of logical analysis.[29]

All that social science could do was to provide, through the advancement of predictions, more or less effective criteria for the ascertainment or reduction (but never removal) of the risks of action taken in conditions of uncertainty.

The reason why predictions and imperatives were so frequently joined together lay, Neurath observed, in the fact that they were both oriented towards the future. But there was a deep difference between them in that commanding was a form of social behaviour which made special use of language with a view towards a future result. This special use removed prescriptive propositions from the class of theoretical propositions whose job was precisely not to prescribe but to predict. Prediction was instead a proposition, relative to certain phenomena, which one assumed would be consistent with future propositions, relative to other phenomena.[30] Thus the use of linguistic statements did not set out directly to produce effects by influencing or restricting the behaviour of others, but intended merely to provide information which could be taken as the basis for a decision. Theory did not make decisions for others, but offered others the elements of knowledge by which to make a free choice: there were therefore two uses of linguistic statements – informative/predictive and suggestive/prescriptive – which were in principle incommensurable and had no logical continuity between them: "science therefore is one thing, practical decision-making another".[31]

Despite the fact that he was often called upon to do so, the social scientist could not decide for humanity.[32] As a specialist in prediction and social planning he could only present a range of possibilities, all equally plausible from a scientific point of view. It would fall to those responsible for political decisons to determine which possibility to realise amongst the plurality of scientific indications and predictions offered them. The social scientist's position with regard to political decisions was analogous to that of an expert in probability calculus asked for an opinion on the chances of winning and losing in different games or different types of gambling.[33] The expert could provide the player with precise information on the structure of each game and the logical calculation of chances involved in each bet, but he could not take decisions for him (by suggesting, for example, that he should pursue a policy of *maximin*) nor would he be able, even if so requested, to provide a criterion which would enable him to win under any circumstances. A player who wanted this would have to turn to an astrologer rather than a mathematician. Similarly the politician who wanted univocal evidence, fixed criteria, and infallible predictions to work on would have to turn not to the social scientist but to a prophet or a political messiah fired by belief in the existence of 'iron laws of history'.[34]

On the basis of these general epistemological premises Neurath went on to criticise in particular the Kelsenian idea of jurisprudence as a 'normative science', to which he gave detailed attention both because of the wide cultural influence of Kelsen's thought and because of the proximity it had to the work of the Vienna Circle. Kelsen, as is well known, took part during the Thirties, albeit intermittently, in the Circle's activities along with Felix Kaufmann and Ludwig von Bertalanffy.[35]

Neurath took the view that Kelsen's contradistinction between the *Sein* and the *Sollen*, the first being conceived of as the realm of 'facts' and the second as the realm of 'values' was devoid of any empirical relevance. In his *Allgemeine Staatslehre*, published in Berlin in 1925, Kelsen had argued that the State and legal system, in so far as they were systems of norms and values, did not belong to the realm of nature and psycho-physical relations but to the 'realm of the spirit' (*im Reiche des Geistes*). He had seen politics as 'part of ethics' and as 'moral knowledge', morality having the job of fixing 'objective goals' (*objektive Ziele*) in human behaviour and of rendering certain courses of action obligatory.[36]

Neurath rejected the idea that 'objective goals' could be determined in human behaviour and dismissed Kelsen's Kantian distinction between "the two worlds of the 'ought to be' and of the 'is'"[37] as mere speculative metaphysics. In addition, he went further in exploring the question of the possibility of an empirical and analytical study of law in its twin aspects of social institution and normative system. This led him, in consequence, to tackle also the relation between law and morality and the basic problem of the foundations of social morality.

He did not argue against the usefulness of Kelsen's attempt to establish a 'general theory of law':[38] when law was examined with the aim of "establishing whether a system of norms was logically consistent", the legal philosopher was working correctly and legitimately, in that he was doing nothing other than applying logical categories to the relations between the norms and seeking thereby to work out the logic of prescriptive propositions. "Demonstration that the correlations of certain norms or legal systems are logically equivalent to other correlations, however unclear at first sight", could have an undoubted potential importance from a practical point of view.[39] Equally important might be an outline of the process by which the norms included in legal texts were established or an inquiry into what the social consequences would be of a more or less exact application of the norms.[40] Even when an examination of the apparatus of law aimed to discover the relation between the logical premises behind legal texts and the principles actually followed in legal practice, this type of research was fully justified, even if still formal in character.[41]

Neurath granted that all this had importance for the purpose of the sociological prediction of legal phenomena, but it did not, to his mind, in any way permit the assumption that legal science was an exceptional normative discipline, that it was, as Kelsen claimed, a 'moral science'. Nor did it open the way for the language of law, wich was essentially composed of prescriptive propositions, to be integrated in the language of the empirical sciences. This language, although it contained evaluative elements, was not normative.[42]

In order to became a science, jurisprudence had to adopt the language and methods of all other empirical disciplines: it had to become a sociology of law. This had been the aim of realist philosophers of law such as Felix Cohen and especially of such empiricist legal theorists of the Uppsala school as Alf Ross, whose *Kritik der sogenannten praktischen Erkenntnis* Neurath reviewed favourably in *Erkenntnis*.[43] But an empiricist translation of the theory of law required, as Neurath saw it, so thorough-going a reconsideration of legal language and ideas that it was legitimate to enquire whether it would in the end be able to establish itself as an autonomous discipline within sociology as a whole, and to assert the claims of its place as 'empiricist jurisprudence'. The empirical aspects of legal practice could easily be lost sight of within a multitude of social disciplines: concepts such as 'crime' and 'punishment', widely employed as dogmatic legal terms, might well, even when redefined in a purely empiricist manner, run the danger of remaining 'erratic' in the context of anthropology and social engineering.[44]

In the light of these epistemological premises behind Neurath's analysis of the legal system, it comes as no surprise to find that he went beyond the denial that law could be seen as a part of or extension to a more general set of normative ethics. Any analysis, he found, of moral phenomena which did not end as a historical and sociological inquiry into the belief, moral behaviour, cosmologies, and folklore which popular traditions transmitted from one generation to the next, was irrelevant.[45]

Having dispensed with all hypotheses of a foundation of moral norms in theology or natural law,[46] Neurath went on to discuss the possibility of a rational 'justification' of social ethics and came to a wholly negative conclusion. He referred critically above all to Kant, whose categorical imperative he saw as a theological remnant, the 'laws of God' separated from their subject and constructed as 'laws 'an sich''.[47] Ethics was in origin the discipline which sought to establish the totality of God's commandments, beginning as a body of case histories of actions forbidden by God and as a doctrine of sin. But as soon as one took leave of the idea of God and attempted to move from divine imperatives to the 'categorical imperative', one committed oneself to a *non sequitur*, just as if one spoke of a man who was a 'neighbour *per se*, although without neighbours'.[48] Foundation of a system of normative ethics, Neurath argued, was not possible from an empiricist standpoint: once the archaic religious tradition was stripped of its theological and metaphysical assumptions, all that remained were "propositions on certain forms of behaviour or commands (*Befehle*) given by certain people to certain other people". Ethics, in a modern non-speculative sense, could be conceived of only as a series of 'habits and rules' on which individuals and groups agreed to found the basis of their common life.[49]

In place of the archaic doctrine of imperative ethics it was possible, Neurath argued, for the empiricist standpoint to put a lay, contingent, and positive conception of morality. This would mean a morality shorn of all 'justification' going back to 'eternal values' or to universal ethical categories,[50] a teleological non-deontological morality which would invite individuals to practise tolerance, democracy, and loyalty – together with acceptance of the sacrifices necessarily entailed by these – if they desired to achieve the "possibility of living within a tolerant environment, living in an atmosphere of friendship, security, and scientific freedom", which would enable them to "enrich the pattern of their lives".[51] It would mean a eudaemonistic morality directed towards the individual and collective pursuit of a 'happiness' which would not be of a Platonic or Kantian but of an Epicurean and Marxist kind.[52]

3. HAPPINESS, UTILITARIANISM, AND SOCIAL ENGINEERING

From the very beginning two themes dominated Neurath's ethical and meta-ethical thinking. The first concerned the replacement of traditional theological/deontological morality with an ethical standard of Epicurean-based 'happiness' (which was accompanied also, in Neurath's formulation, by sharp criticism of the classical tradition of ethical utilitarianism from Helvétius to Hollbach and Bentham). The second concerned the ethical and political critique of the market economy as a system which failed to provide any criterion for the distribution of collective resources and whose compliance with the principle of market freedom led it not only to ignore the question of the full use of economic resources and to provide no answer to the problem of its recurrent crises, but also even to theorise on the expediency of unemployment and the deliberate destruction of goods produced.[53]

Against this doctrine, drawing on the 'scientific utopianism' of Walter Rathenau, Joseph Popper-Lynkeus, and Carl Ballod-Atlanticus, and, to a lesser extent, on Marxism, Neurath set a connected series of counter-proposals: first, an administered or planned economy modelled on the 'war economy' and taking 'natural values' as well as prices into its calculations; second, the socialisation of production, distribution, and consumption; third, an economic and political 'social engineering' which would take as its central means of prediction and planning the systematic statistical study of living standards.[54]

Neurath's early works showed a clear appreciation of the delicate problem of the relation between the rationale behind an individual's actions in his personal pursuit of 'happiness' and the hypothetical rationale behind collective choices made in the economic and political, and hence imperative, advancement of 'social happiness'. He failed, however, to elaborate to any real extent an area which remained the most serious gap in the development of his political thought – the political and institutional theme of the relation between individual and collective choices, and the problem of freedom, individual rights, and dissent. In the early Forties, however, towards the end of his life, he showed himself to be well aware of the problem of democracy and political tolerance when, while not changing his stance in the face of Hayekian views of freedom, he turned the main force of his criticism against totalitarian political doctrines and, in advance of Popper, denounced the 'political Platonism' of the National Socialist concept of the State.

In certain early works, especially 'Das Problem des Lustmaximums' of 1912, he had given clear reasons for his adoption of a eudaemonistic view of morality as well as for his rejection of Benthamian utilitarianism. This was a position he reinforced in one of his last works, *Foundations of the Social Sciences*.

In opposition to mathematical economists who concerned themselves exclusively with the theory of prices, circulation of goods, and economic demography – a clear reference to Jevons, Walras, and Pareto – he argued that a strong link needed to be forged between economic theory, in its classical macroeconomic form, and a eudaemonistic conception of morality. Instead of concerning itself with the reductive scheme of the *homo oeconomicus*, the science of economics should take as its central theme the economic conditions of the pursuit of 'happiness', so turning itself into 'felicitology' or 'eubiotics'.[55]

This belief did not, however, lead Neurath to think that economics could set out to tackle and more or less solve the problem of the 'greatest happiness of mankind', as had been the claim of utilitarians from James Mill to John Stuart Mill and Bentham. The maximisation of 'happiness' or well-being could be seen without difficulty as the aim behind the actions of the single individual, but the concept of 'greatest pleasure' became meaningless *(sinnlos)* when referred to the social group.[56] What he did believe could be argued was that the pleasure or well-being of different people could be compared and that profiles (*silhouettes*) of living conditions were capable of being traced and compared with one another. But this comparison could only be of an ordinal and multiple kind, since there was no means possible of measuring 'degrees of pleasure' and comparing them in size on a cardinal scale even by reference to the same type of 'pleasure' and to the same individual.[57] While it was extremely important therefore to achieve an empirical evaluation of the effects which certain political and economic decisions had on the 'happiness' of a social group, the possibility still had to be ruled out that the utilitarian principle of the 'greatest happiness' could be taken as the basis of a moral or legal system or of a whole system of social organisation.

Given the hypothesis that a political decision was bound to have the effect of simultaneously raising the well-being of one group and depressing that of another, it would not be possible to avoid considering each 'eudaemonistic system' (*Glückskonstellation*) as a whole (*als Ganzes*) without the possibility of making further calculations, individual by individual and pleasure by pleasure, by means of cardinal additions and subtractions.[58] A choice would therefore be made as the discretionary outcome of a political decision, the expression of a subjective preference not based on analysis or 'reason'. Nor could it be argued, except on the basis of a highly problematical psychological fiction, that the maker of the political decision could, by calculating the overall future effect of his decision, put himself in the position of a neutral and sympathetic 'observer' (*Beobachter*) of the varying outlooks of the different groups.[59] This procedural substitute (*Ersatz*) – Neurath argued, as if anticipating a reply to contemporary theses of neo-utilitarianism and neo-contractualism[60] – was profoundly suspect and could in no event be adopted as the foundation of a moral or a legal philosophy.[61] Each individual, as member of a social group, was from the eudaemonistic point of view an autonomous unit unrelated to the sum or average of 'others' happiness'.

Anyone claiming therefore to build a complete system of morality (*ein vollständiges System der Moral*) could not avoid working from a particular and subjective standpoint: he had to adopt the stance of the individual moralist, or social legislator or of God.[62] Any ethical system which aimed to control the social relations of a group – and was not simply the moral option of an individual, freely and consciously adopted for his own choices alone – was bound to take on the imperative structure of command and, without any basis other than will or advantage, to elevate a system of purely personal or particular preferences into a general and normative one. This would hold of course only under the obvious condition that the moral unanimity of all the members of a group was not supposed: but this was a hypothesis lacking any sociological interest and one which was in any case alien to the utilitarian philosophical stance.[63]

In point of fact the utilitarian ethic developed a normative standard which it held to be founded in reason and not purely in convention: utilitarianism did not go back to a preliminary consensus between members of a group or proponents of a rule of distribution. On the contrary, its aim was to eliminate or reduce the propensity to social conflict by asserting the claims of a universal rational principle against ethical and political sectionalism and pluralism. But this was not, Neurath argued, an aim which could be realised:

> a social system cannot generally be got off the ground when it gives equal weight to individuals' differing points of view on the best distribution of social benefits.[64]

It was quite impossible to arrive at a rational method of resolving disagreement over the criteria to be used for distribution. On the theoretical level, rather than hammer out some sort of inconsistent metaphysical doctrine – and the utilitarian ethic was one of these – one would do better to decide by drawing lots. On the level of actual practice it was struggle (*der Kampf*) which was "probably going to decide which criterion of distribution would prevail".[65]

Having once dismissed the classical utilitarian position as an attempt to 'justify' a certain normative scheme of the distribution of resources and social advantages, it was not the case, as he appeared to hint in *Das Problem des Lustmaximums*, that Neurath instead espoused a radical and thoroughgoing philosophy of conflict. He did not reduce ethics to economics and to simple competition between different functions of utility in a zero-sum game, however tempered it might be by the agreed acceptance of its rules. He remained loyal to a humanistic position, even during the period when his adherence to Marxism made it feasible for him to use a classist scheme for the adoption of a particularistic stance. Instead he called for economics itself to become a science of 'social happiness' and an empirical pursuit of collective well-being. It was precisely in this that the originality of his connection of Marx with Epicurus lay, together with his view of Marxism as 'a sort of social Epicureanism' (*eine Art Sozialepikureismus*).[66]

Unlike Plato, Epicurus had made philosophy accessible to women and slaves. His aim had not been the attainment of exalted intellectual goals or spiritual purification, but to "teach the greatest possible number of people the way to become happy". This was not an aristocratic teaching, serving an elite or a particular class: the object was to free all men from fear, whether of God, death, moral laws, or even of the tyranny of scientific laws. So too Marxism, in just the same way as Epicureanism, aimed to struggle for "the happiness of men of all classes and of the *whole of the humanity*".[67]

This eudaemonological version of Marxism may well appear somewhat ingenuous today, and to rest on an optimistically simple view of mankind. But its originality should not be overlooked in the context of European and Russian interpretations of Marxism at the time, dominated as they were by Hegelian historicism and popular Darwinism, as well as being strongly anti-individualistic and based on an ethic of manual labour, progress, industrial development, and man's control of nature. Nor should it be forgotten that Neurath's espousal of the socialist position did not extend either as far as historical determinism[68] – by which socialism was seen as the inevitable result of the development of immanent historical laws – or as far as the neo-Kantian versions then current in Vienna which saw socialism as something very close to a traditional theory of social justice.[69]

Neurath quite openly presented his adherence to socialism, as later to a post-Marxist position on social engineering, as a moral option 'without foundations' and as a personal choice, made in conditions of uncertainty, of one 'path of life' from among the many others open to him.[70] It was a path for which he had shown his preference because of the many positive consequences which could be discerned in it,[71] because of its power to influence 'standards of living', and because of the eudaemonistic efficacy it had in enriching the quality of life.[72]

It seems natural to observe at this point that, however original this was as a moral solution, it left quite unresolved the problem of possible conflict between the personal pursuit of (individual) happiness and the social pursuit of (collective) happiness. This was so both from the point of view of the person actively working to promote the political objective of 'collective happiness' and from the point of view of the person who might find himself experiencing an imperative scheme for 'collective happiness'. The first aspect received no real attention in Neurath's thought: with him the socialist element prevailed over the Epicurean, obscuring the 'apolitical' side of Epicurean eudaemonism and forgetting the Epicurean precept to live unnoticed and enjoy the fruits of friendship safely removed from the upsets of political life. It seems that, to Neurath's mind, there were no room in life for a private pursuit of happiness or that he did not perceive there to be any basic incompatibility between the pursuit of personal happiness and well-being and a commitment to achieving the happiness and well-being of everybody else. In this respect Neurath's anthropological view appears to have owed more to Aristotle and Rousseau than it did to Epicurus.

The second aspect, as we have seen, had been given full treatment by Neurath from the time of his earliest writings: no normative model, he had argued, of the distribution of resources and social benefits could be said to be rationally based and to have claim therefore to impose itself on the wishes of individuals. Legitimation of dissent and consideration of social conflict as the only practical solution possible for an antinomy which could not be resolved by theory were the logical concomitants of this, although, as we have indicated, Neurath failed to follow through the philosophical, political, and institutional implications of this meta-ethical position. In all the stances Neurath took – in his eudaemonistic conception of political economics, his early interest in the economics of war, his proposal for a planned peacetime economy, his perpetual depreciation of the social and economic functions of the monetary system, his unbending opposition to the market economy, his schemes for full socialisation (*Vollsozialisierung*) which denied the right to strike and ignored the question of worker's democracy[73] – there is a consistent thread of implicit political consensualism to be discerned.

Despite the theoretical importance of many of his formulations – acknowledged even by such severe critics as Weber and Hayek[74] – Neurath emerges as a 'scientific utopian' and social engineer who paid little attention to the problems of institutional guarantees of individual liberties, or of the pluralistic organisation of the process of political decision-making, or of the constitutional protection of dissenting minorities. For at least twenty years he appears to have put such questions quite out of mind and so, clearly, to have shared the limitations of a whole tradition of thought in political philosophy stretching from enlightenment to socialism and Marxism.

4. PLANNING FOR FREEDOM: NEURATH'S CRITICISM OF POLITICAL PLATONISM AND THE DISPUTE WITH HAYEK

It is much to Neurath's credit that he was able fundamentally to revise his work on political philosophy in the light of actual experience of totalitarian regimes, both National Socialist and Soviet, and to put the more direct contact with Anglo-Saxon democratic traditions which he gained in Oxford in the last five years of his life to immediate use. In this respect his 'International Planning for Freedom', written and published at Oxford in 1942 in *The New Commonwealth Quarterly* whose editorial board he had joined, provides a good example, as do also the shorter articles of his which appeared in various London journals between 1943 and 1945.[75]

Once again Neurath asserted vigorously the need for economic planning as a permanent means of counteracting the dangers of economic crises and unemployment. But in contrasting a 'technically efficient' model of planning – of which he was critical – with a model efficient 'in producing happiness' – of which he approved – he introduced an important new element. Social engineering, he maintained, should take account of the fact that the instruments themselves of planning were not unrelated to the end which the planning aimed to achieve. If the social engineer had no more than 'technical efficiency' in view, he would not hesitate to employ authoritarian political means, such as a rigid hierarchical structure, Draconian measures, suppression of trade union freedom, and a single party political system. This had been shown both in Nazi Germany and in Soviet Russia.[76]

If, on the other hand, a system of planning aimed to achieve social well-being, it would recognise that respect of personal independence, democracy, voluntary cooperation, and the 'joy of self-government and freedom' were all elements which went to make up the quality of life. In which case not merely the contents but also the forms of planning would need to be directed towards the realisation of these values.[77] The American Declaration of Independence saw liberty and the pursuit of happiness as 'inalienable rights'. However difficult it was to establish what significance the recognition of 'inalienable rights' would have in constitutional law, it was beyond doubt – he now argued – that the political institutions of a country, in upholding or suffocating political and civil liberties, had an enormous influence on its citizens' standard of living.[78] Even the maintenance of certain traditions, although they entailed some small inefficiency and waste of resources,[79] should be brought into the eudaemonological calculations of a democratic social engineering. Rather than following technocratic ideology[80] in setting out to 'maximise' efficiency in the various sectors of production and services through heavily restrictive use of authority, democratic planning would aim principally to inform and educate the public, without recourse to the censorship so regularly practised under single party systems.[81]

Apology for the values of civil liberty and tolerance, a pluralistic social system allowing for a wide variety of civil 'loyalties' and commitments, and the adoption of decentralised political institutions could all, Neurath agreed, appear to contradict the demand for a 'rational' and efficient control of social development and might seem a source of disorder and confusion. Democracy, as its critic Plato had pointed out, was bound to produce not only 'the greatest variety of human nature' but also a certain amount of disorder. In the long term, however, it was in the nature of democracy to prove itself technically more efficient. Unlike single party systems it did not make it necessary that "an essential part of all administrative energy is consumed by fighting against deviations"[82] but instead encouraged the formation of self-stabilising social mechanisms in as much as thinking compliance with the demands of authority took the place of the passive obedience engendered by the command of the all-powerful State.[83] Democratic social engineering would in any case guard against measuring freedom in terms of business efficiency and adapting it to a yardstick of the maximisation of economic or technical results.[84]

In opposition to Burnham in *The Managerial Revolution*[85] and Hayek in *The Road to Serfdom*,[86] Neurath argued against the idea that economic planning led *necessarily* to the suppression of freedom and that there could not, *for scientific reasons*, be any alternative between a planned technocratic economy and a market economy. He granted that Hayek – like Hans Kelsen,[87] Werner Fite[88] and R.H.S. Crossman[89] – had admirable reasons for denouncing the diffusion through Europe of ideologies which resurrected the model of a planned State which Plato presented in his *Republic*.[90] There was no doubt in his mind that Plato's idea of the State was not only authoritarian and anti-democratic but even also militarist and racist and that the National Socialist 'planners' had not therefore been mistaken in according it their respect. As a result of this he found himself directly involved in a war of words with such leading experts on Plato's thought as G.C. Field and C.E.M. Joad.[91] Pre-empting Popper's well-known arguments in *The Open Society and Its Enemies*,[92] he advocated a re-reading of Plato's political philosophy which would see in the philosopher kings of the *Republic* the paradigm of a system of political planning in which human individuality was entirely sacrificed to fixed collective roles, and in which the idea of planning was used to bolster the unity and stability of the governing class, their intellectual and racial superiority, and the strict segregation between planners and planned.[93]

Where Hayek erred, however, was in deducing from undeniable occurrences of 'planning' being used as a means to introduce totalitarian leadership and as a vehicle for fascist propaganda that all forms of planning were basically the same and that all were 'roads to serfdom'.[94] There was no scientific reason to prevent construction of the theoretical model of a society in which procurement of raw materials and foodstuffs, for instance, was administered by representative political units democratically responsible for laying down guidelines for the distribution of goods in the light of a comprehensive overview of people's various expectations.

Democratically-based planning of labour and the economy would be entirely compatible with respect for the rights of minorities and with pluralism in economic organisation and the means of information. Central planning units would have their powers of decision strictly limited to general matters (e.g. designation of the extent of the national rail network) and central planners would be unable to exercise totalitarian will through the imposition of uniformity on social choices.[95] Although free competition would be appreciably restricted, this type of planning would bring about a process of social cooperation based on compromise (*Kompromiß*). Hayek, Neurath argued, had not only failed to consider this possibility but had inveighed dogmatically, and with especial harshness, against the very people who were proposing the model of 'planning for freedom'. His argument, conducted purely in general terms and without any persuasive backing, was that the wider a system of planning was, the more impracticable it became, unless accompanied by a commensurate restriction of freedom.[96] It was an argument which, as is well known, was to be taken further and presented in more methodologically radical terms by Popper in *The Open Society and Its Enemies*: in the place of holistic and irrational 'utopian engineering' Popper was to set a wholly rational 'gradualistic social engineering' compatible with the 'scientific method' and the protection of political liberties.[97]

Neurath objected that the idea of freedom from which Hayek started was substantially anti-democratic, in that he conceived democracy and the principle of majority rule as entailing the permanent danger of a totalitarian breach of trust on the part of despotic majorities when faced with dissenting minorities. Furthermore Hayek's enthusiasm for the market economy and free competition led him to jettison all consideration of social value: he ignored the way in which the market seriously malfunctioned not only as a mechanism for the allocation of wealth but also as a system of utilising resources and productive energies; the problems of poverty and unemployment passed him wholly by. It was in place of this unyielding form of anti-democratic protoliberalism that Neurath proposed to set his idea of democracy as 'the acknowledgment of nonconformism within a freely accepted social order'.[98]

Democratic control of the processes of production and distribution started from the assumption that freedom, unless it was to be understood as an intangible privilege of a minority, could not avoid being socially 'produced':[99] it could not be conceived of as a spontaneous result of social interaction but should rather be thought of and intended as the outcome of a conscious intervention on the part of society in its own processes of development. Planning for freedom, i.e. programming freedom politically, was hardly 'anti-scientific'; it was in fact the only way an empirical sociology could see as being open to it when its objective was not the jealous protection of the freedom of a few but the extension of freedom as an essential element in the quality of life.

Unilateral and dogmatic subscription to the principles of free market economics ran counter, in just the same way as a totalitarian system of planning did, to the experience of modern empiricism which had found that no single best solution existed to social problems, but that a multitude of further possibilities and intermediate solutions always lay open.[100] In this case too empiricism should lead us to adopt an attitude of prudent scepticism, distrusting the certainties and narrow alternatives offered by 'pseudorationalism'.[101]

NOTES

[1] O. Neurath, *Empirische Soziologie*, [1931e], *GpmS*, p. 516, *ES*, p. 406 ("Theorie ist eben auch nur eine Form der Praxis, ein Ausschnitt aus ihr. [...] Noch mehr gilt das von der Soziologie. Der Gelehrte ist ein Element wie jedes andere").
[2] Ibid., loc. cit.; O. Neurath, *Foundations*, [1944a], pp. 46–7.
[3] Ibid., p. 43.
[4] O. Neurath, *Empirische Soziologie*, [1931e], *GpmS*, p. 516, *ES*, p. 406; O. Neurath, 'Wissen und Sein', [1938a], *GpmS*, p. 841.
[5] Ibid., loc. cit.
[6] O. Neurath, *Foundations*, [1944a], p. 42.
[7] Ibid., loc. cit.
[8] Ibid., loc. cit.
[9] Ibid., p. 43.
[10] Ibid., loc. cit.
[11] Ibid., loc. cit.
[12] O. Neurath, 'Prediction and Induction', [1946a], p. 1, *PP*, p. 243.
[13] O. Neurath, *Foundations*, [1944a], pp. 43, 45–6.
[14] O. Neurath, 'Prediction and Induction', [1946a], p. 1, *PP*, p. 243; O. Neurath, *Foundations*, [1944a], p. 46.
[15] Ibid., p. 44; O. Neurath, *Empirische Soziologie*, [1931e], *GpmS*, p. 526, *ES*, pp. 418–9.
[16] O. Neurath, *Foundations*, [1944a], p. 43.
[17] Ibid., loc. cit.; O. Neurath, *Empirische Soziologie*, [1931e], *GpmS*, p. 526, *ES*, p. 418.
[18] O. Neurath, *Foundations*, [1944a], p. 43; cf. G. Statera, *Logica, linguaggio e sociologia. Studio su Otto Neurath e il neopositivismo*, pp. 174–5.
[19] O. Neurath, *Foundations*, [1944a], pp. 45–6.
[20] Ibid., p. 41.
[21] Ibid., pp. 46, 10, 41, 38.
[22] O. Neurath, *Foundations*, [1944a], p. 38.
[23] Ibid., p. 44.
[24] O. Neurath, 'Physicalism: The Philosophy of the Viennese Circle', [1931c], p. 622, *PP*, p. 50.
[25] O. Neurath, 'Zur Theorie der Sozialwissenschaften', [1910], *GpmS*, p. 41.
[26] Ibid., pp. 41–2.
[27] O. Neurath, 'Soziologie im Physikalismus', [1931f], p. 430, *PP*, p. 89.

28 'A' refers to a theoretical hypothesis, 'B' to the initial conditions of the analysed phenomenon, and 'C' to a probable class of events.

29 O. Neurath, 'Soziologie im Physikalismus', [1931f], p. 430, *PP*, p. 89.

30 Ibid., loc. cit.

31 O. Neurath, *Empirische Soziologie*, [1931e], *GpmS*, p. 527, *ES*, p. 419.

32 O. Neurath, *Foundations*, [1944a], pp. 43–4.

33 Ibid., p. 44.

34 Ibid., loc. cit.

35 Kelsen published two essays in *Erkenntnis*: 'Die Entstehung des Kausalgesetzes aus dem Vergeltungsprinzip', *8* (1939), pp. 69–130; 'Causality and Retribution', ibid., Appendix, 1939–40, pp. 424–9. In 1946, but bearing the date of 1941, Kelsen's essay 'Vergeltung und Kausalität' appeared in the series 'Library of Unified Science', which was edited by Neurath. The English translation of this essay had already appeared in 1943: *Society and Nature*, Chicago: University of Chicago Press. References to Kelsen, Kaufmann and von Bertalanffy's participation in the activities of the Vienna Circle can be found in F. Barone, *Il neopositivismo logico*, pp. 39, 352, 687–8.

36 Cf. H. Kelsen, *Allgemeine Staatslehre*, Berlin: Verlag Julius Springer, 1925, pp. 14, 27. Cf. also O. Neurath, 'Soziologie im Physikalismus', [1931f], *GpmS*, pp. 420–1, *PP*, p. 81.

37 O. Neurath, *Foundations*, [1944a], pp. 40–1; O. Neurath, *Empirische Soziologie*, [1931e], *GpmS*, pp. 501–2, *ES*, p. 390; O. Neurath, 'The Social Sciences and Unified Science', [1939/40], p. 433, note 7, *PP*, p. 212, note 8. Neurath also criticised Felix Kaufmann who maintained the analogous opposition between 'internal world' and 'external world' and attributed the realm of law to the internal world. Cf.: ibid., loc. cit.; O. Neurath, *Foundations*, [1944a], p. 50, note 51.

38 O. Neurath, 'Soziologie im Physikalismus', [1931f], p. 421, *PP*, p. 81.

39 Ibid., pp. 420, 421, *PP*, pp. 80, 81.

40 O. Neurath, 'Soziologische Prognosen', [1936h], p. 403.

41 O. Neurath, 'Soziologie im Physikalismus', [1931f], pp. 420, 421, *PP*, pp. 80, 81.

42 O. Neurath, 'Soziologische Prognosen', [1936h], p. 403.

43 O. Neurath, Review of A. Ross, *Kritik der sogenannten praktischen Erkenntnis* (Leipzig: Felix Meiner, 1933), *Erkenntnis*, *5* (1935), pp. 366–7.

44 O. Neurath, *Foundations*, [1944a], p. 41; O. Neurath, 'Mensch und Gesellschaft in der Wissenschaft', [1936b], *GpmS*, p. 717.

45 O. Neurath, *Foundations*, [1944a], pp. 41, 11.

46 O. Neurath, 'The Social Sciences and Unified Science', [1939/40], p. 431, *PP*, p. 210; O. Neurath, 'Universal Jargon and Terminology', [1941], pp. 139–41, *PP*, pp. 222–4.

47 O. Neurath, 'Physicalism: The Philosophy of the Viennese Circle', p. 622, [1931c], *PP*, p. 50; O. Neurath, *Lebensgestaltung und Klassenkampf*, [1928], *GpmS*, pp. 280–7, *ES*, pp. 282–90.

48 O. Neurath, 'Soziologie im Physikalismus', [1931f], p. 418, *PP*, p. 79.

49 O. Neurath, *Empirische Soziologie*, [1931e], *GpmS*, pp. 526–7, *ES*, pp. 418–9.

50 O. Neurath, *Foundations*, [1944a], p. 44.

51 Ibid., loc. cit.

52 O. Neurath, *Lebensgestaltung und Klassenkampf*, [1928], *GpmS*, pp. 280–7, *ES*,

pp. 282–90. Cf. G. Statera, *Logica, linguaggio e sociologia. Studio su Otto Neurath e il neopositivismo*, pp. 178–83.

[53] O. Neurath, *Foundations*, [1944a], p. 35; O. Neurath, 'International Planning for Freedom', *The New Commonwealth Quarterly*, I, April 1942, pp. 285–6; II, July 1942, p. 23, now also in *ES*, pp. 425–6, 431–2.

[54] O. Neurath, *Foundations*, [1944a], p. 37. On this subject see E. Nemeth, 'Die Einheit der Planwirtschaft und die Einheit der Wissenschaft', in R. Haller (ed.), *Schlick und Neurath. Ein Symposion*, pp. 437–49.

[55] O. Neurath, 'Zur Theorie der Sozialwissenschaften', [1910], *GpmS*, pp. 31–2, 43; O. Neurath, 'Das Begriffsgebäude der Wirtschaftslehre und seine Grundlagen', *Zeitschrift für die gesamte Staatswissenschaft, 73* (1917), now in *GpmS*, pp. 104–5, 106–7; O. Neurath, 'Das Problem des Lustmaximums', *Jahrbuch der Philosophischen Gesellschaft an der Universität zu Wien 1912*, now in *GpmS*, pp. 47–8; O. Neurath, *Foundations*, [1944a], pp. 37, 39–40.

[56] O. Neurath, 'Das Problem des Lustmaximums', [1912], *GpmS*, pp. 47–8.

[57] Ibid., pp. 51, 53; O. Neurath, 'International Planning for Freedom', I, [1942], p. 286, *ES*, p. 426.

[58] O. Neurath, 'Das Problem des Lustmaximums', [1912], *GpmS*, pp. 52–3.

[59] Ibid., pp. 53, 54.

[60] Cf.: J.C. Harsanyi, *Rational Behavior and Bargaining Equilibrium in Games and Social Situations*, Cambridge: Cambridge University Press, 1977; J. Rawls, *A Theory of Justice*, Cambridge (Mass.): Harvard University Press, 1971.

[61] O. Neurath, 'Das Problem des Lustmaximums', [1912], *GpmS*, p. 54.

[62] Ibid., loc. cit.

[63] Ibid., loc. cit.

[64] Ibid., p. 55.

[65] O. Neurath, 'Das Problem des Lustmaximums', [1912], *GpmS*, p. 55.

[66] O. Neurath, *Lebensgestaltung und Klassenkampf*, [1928], *GpmS*, p. 286, *ES*, p. 289.

[67] Ibid., loc. cit.

[68] Ibid., *GpmS*, p. 289, *ES*, p. 293.

[69] According to Neurath Marxism could not be understood as a theory of justice, cf. O. Neurath, *Empirische Soziologie*, [1931e], *GpmS*, pp. 449, 454–5, *ES*, pp. 346, 353.

[70] O. Neurath, *Foundations*, [1944a], p. 44.

[71] O. Neurath, *Empirische Soziologie*, [1931e], *GpmS*, p. 526, *ES*, pp. 418–9.

[72] O. Neurath, *Foundations*, [1944a], pp. 35–6, 44.

[73] Cf. R. Hegselmann, 'Otto Neurath. Empiristischer Aufklärer und Sozialreformer', pp. 19–23, 23–33.

[74] Cf.: F.A. Hayek, *The Counter-Revolution of Science*, pp. 45, 96, 219; F.A. Hayek, *Collectivist Economic Planning*, pp. 30–1.

[75] O. Neurath, 'Planning or Managerial Revolution?', *The New Commonwealth Quarterly*, April 1943, pp. 148–54; O. Neurath, 'Ways of Life in a World Community', *The London Quarterly of World Affairs*, July 1944, pp. 29–32; O. Neurath and J.A. Lauwerys, 'Nazi Text-books and the Future', *The Journal of Education, 76* (1944), I, no. 904, pp. 521–2; II, no. 905, pp. 574–6; O. Neurath and J.A. Lauwerys, 'Plato's Republic and German Education', ibid., *77* (1945), no. 907, pp. 57–9; no. 910, pp. 222–4; no. 913, p. 394; O. Neurath, 'Germany's Education and Democracy',

ibid., *77* (1945), no. 912, pp. 370–1.
[76] O. Neurath, 'International Planning for Freedom', II, [1942], pp. 25–6, *ES*, p. 436.
[77] Ibid., I, pp. 283, 287, *ES*, pp. 423, 427; O. Neurath, *Foundations*, [1944a], p. 35.
[78] O. Neurath, 'International Planning for Freedom', I, [1942], pp. 283, 287, *ES*, pp. 423, 427.
[79] Ibid., I, p. 287, *ES*, p. 427.
[80] Ibid., I, p. 287, *ES*, pp. 426–7.
[81] Ibid., I, p. 287, *ES*, p. 427.
[82] Ibid., I, p. 291, *ES*, p. 430.
[83] Ibid., I, pp. 291–2, and II, p. 25, *ES*, pp. 430–1, 436.
[84] Ibid., II, p. 28, *ES*, p. 440; O. Neurath, 'Planning or Managerial Revolution?', [1943], pp. 150–1.
[85] Ibid., pp. 149–54.
[86] O. Neurath, Review of F.A. Hayek, *The Road to Serfdom* (London: George Routledge & Sons, 1944), in *The London Quarterly of World Affairs*, January 1945, pp. 121–2.
[87] Cf. H. Kelsen, 'Platonic Love', *The American Imago, 3* (1942), pp. 1–112.
[88] Cf. W. Fite, *The Platonic Legend*, New York and London: C. Scribner's Sons, 1934.
[89] Cf. R.H.S. Crossman, *Plato Today*, London: Allen & Unwin, 1937.
[90] O. Neurath, Review of F.A. Hayek, *The Road to Serfdom*, [1945], p. 121.
[91] O. Neurath and J.A. Lauwerys, 'Plato's Republic and German Education', [1945a], n. 910, pp. 222–4.
[92] Cf. K.R. Popper, *The Open Society and Its Enemies*, pp. 86–168. In his 'Reply to a Critic', published as 'Addendum III' in the first volume of the fourth English edition of *The Open Society and Its Enemies* (1962, pp. 323–43), Popper replied to R.B. Levinson's book *In Defense of Plato* (Cambridge, Ma.: Harvard University Press, 1953). Popper expressed his surprise that Levinson, in defending Plato's philosophy against its critics, associated Popper himself "with an older compatriot of Popper's, the late versatile Austrian philosopher and sociologist, Otto Neurath". Rather curtly he replied: "in fact neither Neurath nor I had any sympathy for the other's philosophy, as emerges only too clearly from Neurath's and my own writings; Neurath, for example, defended Hegel, and attacked both Kantianism and my own praise of Kant. Of Neurath's attack on Plato I heard for the first time when I read about it in Professor Levinson's book; and I have not yet seen Neurath's relevant papers" (p. 324). Cf. B. Levinson, op. cit., p. 441.
[93] O. Neurath and J.A. Lauwerys, 'Plato's Republic and German Education', [1945a], no. 907, pp. 57–8.
[94] O. Neurath, Review of F.A. Hayek, *The Road to Serfdom*, [1945], p. 121.
[95] Ibid., loc. cit.
[96] Ibid., loc. cit.
[97] Cf. K.R. Popper, *The Open Society and Its Enemies*, pp. 157–201, passim. It is worth recalling Popper's surprising mistake on this subject. In note 4 of chapter 9 (p. 285) Popper attributes to Hayek the idea of a 'planning for freedom' which in fact had been maintained by Neurath against Hayek. In somewhat tortuous manner Popper declares his agreement with Hayek's (supposed) idea. Ironically this is the only case in which Popper (unconsciously) agreed with a political thesis of Neurath's.

[98] O. Neurath, Review of F.A. Hayek, *The Road to Serfdom*, [1945], p. 121.
[99] O. Neurath, 'International Planning for Freedom', II, [1942], p. 28, *ES*, p. 440.
[100] O. Neurath, Review of F.A. Hayek, *The Road to Serfdom*, [1945], p. 122.
[101] Ibid., loc. cit.

CONCLUSION

REFLEXIVE EPISTEMOLOGY AND SOCIAL COMPLEXITY

The interpretation of Neurath's thought which has been advanced in the preceding pages is of course only one of the many which it is possible to envisage. The pluralistic lesson of the *Foundations of the Social Sciences* would be lost if we were not to recognise the possibility of a number of Neuraths, or of 'pluri-Neurath', as he might put it himself. Nor should I, as author, attempt to conceal or minimise the force of the theoretical prejudices which must, inevitably, have contributed both to my selection of topics and to the presentation I have given them within a scheme of discourse for which I alone must bear the responsibility.

Scholarly rigour has naturally been one aim which I have attempted to pursue, so far as I have been able, in my reconstruction of Neurath's thought. Also, and perhaps more importantly, there have been the 'general theses', both historiographical and more properly philosophical, which the re-reading of Neurath's works and reflection on his mind and influence have prompted me to put forward.

As far as the historiographical thesis is concerned, both with regard to its more specific aims concerning the Duhem-Neurath-Quine relationship and with regard to its more general and far-reaching aims – in particular that the 'revolt against positivism' of the Seventies turns out to have been no more than an 'imaginary revolution' – here I hope to have provided ample evidence in my work for the views I have expressed. In some matters, in particular the link between Neurath's conventionalism and Quine's reform of neo-empiricism, or the 'regressive' nature of Popper's falsificationism in the light of Neurath's critique of 'pseudorationalism', I believe that the evidence I have drawn from the works themselves will serve, when taken together with the historiographical arguments I have adduced, to confirm the theses I have advanced.

With regard also to the association of Carnap and Neurath, I believe I have shown that there are strong arguments for seeing this as an association which arose from a variety of practical circumstances and a significant number of philosophical misunderstandings. Even in the International Encyclopedia of Unified Science, which was conceived by each of them in entirely different ways, their partnership was far from being a true convergence of philosophical aims and scientific projects. Convincing confirmation of this, emphasised by their dramatic quarrel in 1943–45, comes to light in the Neurath-Carnap correspondence.

The more general historiographical aspects which I believe my work points to are of course more open to question and I should not be surprised to see them disputed in future. But the philosophical and historiographical questions which I have raised, however differently they may be weighed or put in relation to the texts I have analysed, should not now be ignored in the way they have so far, as a result mainly of the influence of the Popperian school (Popper himself, Lakatos, Watkins, and others) and even also of the Carnapian school, to which Hempel too may in this respect be said to belong.

As to the more properly philosophical aspects of my research – which concerns the themes, especially, of 'circularity' and 'reflexivity' in the epistemology of the natural and social sciences – here I believe that my reading of Neurath's work establishes a number of basic positions which it may be helpful for me to set out briefly at this point, without further reference to the sources or to other interpretations of Neurath's thought.

These positions, it cannot be denied, go beyond Neurath's own epistemological view-point and relate it to problems in contemporary philosophy of science of which Neurath himself could obviously have had no knowledge. And yet there can be no doubt that these are positions which all refer either explicitly or implicitly to specific elements contained in Neurath's philosophical thought.

Contemporary philosophy of science seems to me to be marked out now, at the end of the Eighties, by at least the following characteristics:

1. a crisis in the 'received view' of empiricism together with an abandonment of positivist realism and the correspondence theory of truth (despite certain attempts to establish a place for 'non-naive realism');[1]
2. a widespread scepticism about the possibility of constructing a universal scientific language and, consequently, about the value of philosophies of science of such a rationalistic or normative character as those favoured by Carnap, Hempel, or Popper;[2]
3. a large philosophical movement – the so-called 'sociological turn' – in the area of the sociology of science and a renewed, more radical, *Wissenssoziologie*, which, to me, represents the most interesting and promising aspect of 'post-Kuhnean' developments in Western philosophy of science;[3]
4. numerous attempts to overcome the crisis in the 'received view' of

empiricism by means of naturalistic 'short-cuts' which, by resurrecting the neo-positivist ostracism of philosophy (and history), set out to include epistemology within empirical research itself. In accordance with the model of Quine's 'naturalised epistemology', the reflexivity of epistemological analysis becomes submerged within the terms of cognitive neurology, experimental epistemology, evolutionary epistemology, second order cybernetics, artificial intelligence, causalistic *Wissenssoziologie*[4] etc.;

5. a persistent uncertainty over the epistemological standing of the social sciences, which on the other hand now show less of the influence of dogmatic empiricism and behaviourism – here the 'tragedy of political science' comes especially to mind[5] – and the emergence of new problematical areas arising from the increasing complexity of the 'post-industrial' societies which are marked out by the growth of information, robotic and multi-media technologies.[6]

1. REFLEXIVE EPISTEMOLOGY

'Reflexive epistemology' is the term which I propose in order to express the attempt to take further the fundamental requirements of the tradition of non-dogmatic, conventionalistic, pluralistic, and gnoseologically sceptical empiricism within a context of the Neurathian themes of 'reflexivity' and 'complexity'.[7] This attempt may be viewed as a general alternative to 'naturalised epistemologies'.

In 'reflexive epistemology' it is important above all to emphasise that methodological decisions, through which one or more linguistic propositions are stipulatively assumed as the premise of a certain scientific discourse, do not play a part only in the reduction of the theoretical pluralism which is present at every level of scientific abstraction. They take part, even before this, in the assumption as non-problematical of the results of observation and perceptual experience. They are inseparable from the 'problems' considered important by the scientific community which subscribes to the relevant linguistic conventions, and are thus circularly connected to that community's 'expectations' in terms of conceptual habits, philosophical traditions, manipulative techniques, shared social values, disciplinary identities, investments of collective resources. The 'underdetermination' which Duhem argued for in relation to theories therefore also involves 'observation propositions'. Thus it is necessary to recognise that it is not only theories which can be abandoned when they are in conflict with empirical observation: under certain conditions the results of empirical observation also can be – and often are – abandoned if they conflict with certain theories. This principle of symmetry between observation propositions and theoretical propositions has been rightly called the 'Neurath-principle'.

The language of science is essentially no different from ordinary language, at least in the sense that it cannot be systematically separated from the language of everyday speech. Attempts to realise this semantic divide by the analytical 'construction' of rigorous linguistic systems – in Russell as in Carnap, Tarski or, more recently, Paul Lorenzen – run counter to the paradoxes of logical self-reference, unpredicative sentences, and the unavoidable circularity of any definition of 'analyticity'.[8]

The language of science contains therefore the essential characteristics of any language: imprecision, ambiguity, historical variability, psychological contingency, sociological conditioning.[9] All scientific communities are located within semantic circuits in which the contextuality of meanings is a vital condition of the possibility of communication. And every scientific discipline locates itself within linguistic, theoretical, and more general philosophical structures, in order to achieve historical and conventional delimitation of its linguistic ambit, its object, and its procedures, and to achieve stipulative assumption of 'background knowledge' as non-problematical. In this sense, because every change of 'paradigm' or 'encyclopedia' develops circularly from a widespread acceptance of preceding conceptualisations, no scientific revolution is strictly possible if by this is meant, as in the early Kuhn of *The Structure of Scientific Revolutions*, a radical discontinuity or total incommensurability between philosophical paradigms, either contemporary or subsequent.[10] The imprecision of language, as Neurath maintained following Ajdukiewicz, does after all guarantee a possibility of communication.

In order to evaluate the cognitive significance of any scientific work, it is crucial to refer it to the specific 'community' – together with the historical and sociological conditioning of that community – which has produced it. Crucial also is socio-anthropological research on the general contexts which condition the praxis of scientific communities – their values, interests, hopes, fears, myths, semantic rituals, and 'styles of thought' – in just the same way as for any other organised group from a 'primitive' tribe to NASA.

Seen in this way, therefore, even logic and mathematics are to some extent 'empirical sciences'[11] and there is no area of scientific language for which a stability criterion can hold which is not circular, a *noli me tangere* which may be thought of as a nucleus resistant to any refutation or to any meaning variance. Whenever a theorem is examined, whenever the acceptance or refutation of an empirical prediction is at stake, there is also at stake, in the last analysis, the entire apparatus of assumptions, concepts, and theories of science. And this is the thesis which ought by now to be known as the 'Duhem-Neurath' (rather than Duhem-Quine) 'holistic-reflexive thesis'.

Scientific research cannot establish any objective truth of an extralinguistic kind, nor on the other hand can it set out to challenge one, as, in opposite but parallel ways, do philosophies of knowledge of a realistic and idealistic type. Such 'pseudorationalistic' philosophies fail to recognise the situation of cognitive circularity intrinsic to all scientific discourse, and they establish linear, 'directional', relationships with their objects. They 'naively' imagine relations of intellectual mirroring of 'reality' or, conversely, relations of intellectual production of 'reality'.

Reflexive epistemology sees scientific knowledge as being in a position to respond to the situation of circularity by including itself reflexively in the horizon of knowledge, but without ever tracing the complete circle of cognitive self-transparency, by neutralising, so to speak, the anthropological, semantic, historical, and sociological preconditions of the very conditions of departure. Even though endowed with a possibly high level of reflexivity and self-criticism, scientific knowledge is still not able to set itself in a position of unbiased *tabula rasa* from which to set out again in Cartesian manner towards the building – no longer reflexive but directional – of a fixed and objective knowledge. Since it is in the nature of reflexive epistemology to be partial, provisional, and 'relative', it does not attempt to pose as an 'epistemology of epistemology', i. e. as a 'pseudorationalistic' metaphysics of knowledge.

The historico-sociological approach to the phenomenon of science does not allow any justification of relativism which is not itself circularly included in the relativistic view-point of all complex forms of human experience and knowledge. Far from weakening relativism, the recognition of its circularity makes it the point of departure for every epistemological pursuit which is fully conscious of the complexity of the cognitive processes. It rules out not only the 'naive realism' of the positivists but also any kind of 'naive idealism'.[12]

This is, according to my non-reductive and non-naturalistic interpretation, the central meaning of Neurath's nautical metaphor, a meaning which appears to have eluded even Quine, as well as such otherwise acute recent commentators as Paul Lorenzen and Christopher Cherniak.[13]

Scientific research is able therefore to provide only practical precepts of a hypothetical nature for the comprehension and reduction (but never suppression) of the complexity of the environment, and thus for the weighing of the risks involved in every decision taken and every action initiated in conditions of uncertainty. The impressive success our century has achieved through the technological application of certain natural sciences is proof of the high level of practical trustworthiness and the adequacy (however approximate) of the predictions, procedures, and operative techniques which have been gathered within one particular cultural tradition – that of the West – creating an effective core of local rationality. This cultural tradition consequently enjoys both an intellectual advantage and a technological, economic, and military hegemony which are probably unassailable, but this does not prove that this culture commands cognitive instruments which make it more capable than others of approaching the 'truth', if, following Tarski and Popper, this term is taken to mean a metalinguistic correspondence between language and 'reality'.

The progress of science appears in fact to respond not to a criterion of convergence and cumulativity of theories, but to a 'pragmatic criterion'. It is an instrumental progress in terms of a growing capacity for technological prediction within the bounds of a specialised and piecemeal rationality, as is showed, for example, by the permanent and probably increasing difficulties which all attempts to unify theoretical physics have met with. Also it is a progress in the potential (but only potential!) capacity of *homo sapiens* to extend the range of controllability of his environment. But to this 'pragmatic' progress (hindered, however, and often compromised by dramatic side-effects) there does not correspond either a linear historical growth of the conceptual tools of science, or a logical 'transitivity' of scientific propositions, or an asymptotic approximation of them to the 'truth'.[14]

This growth in areas of instrumental knowledge increases the span and complexity of the mental horizon of *homo sapiens*. His 'ignorance' lessens in this way, but his 'certainties' do not increase. On the contrary, the wider the universe of intellectually conceived possibilities becomes, the more complex and uncertain do the theoretical and practical horizons of human experience become. The body of available philosophical knowledge encounters increasing problems in orienting scientific research within the widened realms of astrophysics and microphysics whose comprehension and manipulation appear to require constant changes of paradigm. Almost every category of classical science – space, time, matter, energy, causality, etc. – have undergone and continue to undergo violent changes in contemporary developments of physics. And even the criteria of practical choices – in economics, politics, ethics, law – appear to become ever more contingent and uncertain, ever more 'opportunistic', in the absence of a 'general meaning' which can be attributed to the 'reduction of complexity'.[15]

Empiricism acts, with respect to methodological decisions, as a requirement of pragmatic utilisability of theoretical hypotheses, as an *organon* of prediction, not as an instrument of 'explanation' on the basis of universal laws and as a guarantee of the objectivity of knowledge. If by scientific explanation is meant the presentation of the reasons by which one certain event, and not another, has happened and had to be expected, then it is clear that the empirical sciences cannot offer any adequate explanation, because there exists no possibility of explaining in this sense. Every scientific 'explanation' makes reference to laws which can be assumed as valid only within limited domains and within finite temporal sections, within contingent 'cosmic aggregations', to use Neurath's words. Predictions based on these laws can be considered valid only *ceteris paribus* or *rebus sic stantibus* and are thus subject to the condition of clauses which cannot ever be rigorously defined. Induction always proceeds therefore from particular to particular.[16]

All of this renders unserviceable not only the causalistic scheme of scientific explanation which was proposed by Popper and criticised by Neurath, but also the reduced causalistic model – the deductive-nomological model or covering-laws-model of explanation – elaborated by Hempel in the footsteps of Popper. To predict means to work according to inductive-statistical procedures where extrapolation plays the part of the reduction of complexity by means of the selection of regularities which it is decided to assume as the basis of prediction, after having connected them with one another within a certain theoretical model. As Mach earlier argued, the search for regularities and the formulation of general laws plays a practical role of the 'restriction of expectations' and 'diminution of the possibilities'. The choice of reference classes of statistical generalisation does not underlie any logico-methodological rule and no 'inductive logic' such as Reichenbach's or Carnap's is conceivable.[17] Statistical generalisation obviously does not allow the prediction of 'individual events'. The uncertainty of empirical prediction is moreover confirmed by the rough nature of empirical measurement and by the situation of circularity in which all attempts to realise perfect measurements end, as the uncertainty principle seems definitively to prove.[18]

From the standpoint of reflexive epistemology it is not possible to find any criterion of rigorous demarcation, of a logico-methodological or linguistic nature, between science and metaphysics. If metaphysics is taken to mean a selective and regulating scheme which is not in a position to base itself empirically, then all science is, in this sense, metaphysics. And if by metaphysics is meant a philosophical complex which, for the purposes of prediction, elaborates models, metaphors, and analogies which are not deducible and not testable, except partially, on the basis of empirical experience, then all science is, in this sense too, metaphysics.[19]

A 'rational' reconstruction of the development of science cannot therefore claim to define science, as opposed to metaphysics, as a language or a complex of procedures rigorously identified by a 'method' or a 'logic of scientific research'. Because there is no logico-methodological or linguistic rule available either for the empirical sciences or for metaphysics, the distinction between what should be meant by science and what should be excluded from the realm of science cannot do other than trust in practical criteria of a historical or sociological and essentially non-normative nature. Here too investigation into the values, ideologies, and 'styles of thought' of the scientific communities will be necessary.

Any other approach leads inevitably to an *epistemologia imaginabilis*, as Fleck would say,[20] or to 'pseudorationalism' as Neurath would put it. It leads to epistemologies which superimpose an abstract and absolute model of rationality on the practice of the scientific communities, but without succeeding in providing workers in science with any methodological directions which are at all usable.

2. ONTOLOGICAL AND DEONTOLOGICAL FALLACIES

An epistemological discrimination can be drawn very sharply, however, against the ontological fallacy involved in any attempt to prove linguistically the 'existence' or 'objective reality' of physical bodies or of deities, or that involved in any proposal – such as in Thomas Aquinas, Russell, or Popper's late theorising on truthlikeness and 'world 3' – to assume the 'objective' reference of metaphors as something which precedes the symbolic process of the production of metaphors, or as 'objective knowledge'. In this way it is possible to distinguish a scientific use from a non-scientific, purely speculative, use of metaphysics and to exclude as fallacious realistic ontology (and dogmatic theology) from the realm of scientific knowledge.[21]

This discrimination is equally potent in dealing with the 'deontological fallacies' of all forms of ethical cognitivism and legal or political moralism.

The fact, which is widely accepted in non-dogmatic empiricism, that there is always an evaluative element present in the language of theory and that this is not in the great majority of cases analysable is not sufficient to justify an assimilation between the prescriptive language of ethics (or law) and the theoretical language used in predictions. On the contrary, this fact gives a further and decisive reason for rejecting the deontological perspective of ethical cognitivism. Only on the basis of a realistic ontology which sees knowledge as an intellectual mirroring of reality, as a linguistic correspondence to it, or as a discovery of the objective laws of nature, is it possible to believe that the 'ought to be' of a normative ethics can be established on the basis of the objective 'being' of the world; it is possible, that is, to claim to deduce a deontology from an underlying ontology, by extracting imperatives from assertions, and prescriptive propositions from observation propositions.

Claims to establish rigorously (or scientifically, or cognitively) the validity of a normative system never fail to refer to the distinction between 'facts' and 'values', by which the facts, ascertained by knowledge, are made the basis for the construction of a theory of values. This theory then asserts a claim to assume prescriptive authority, since it presents itself as a deontology established not on the contingency of subjective preferences or of particular traditions, but on the rationality of a natural order. The ethics of natural law preached by the Roman Catholic church with regard to sexuality forms a typical example of this.

In reflexive epistemology the opposition between 'facts' and 'values' would need to give way to the distinction between theories and prescriptions, on the basis of the assumption that the language of theory excludes prescription but necessarily includes evaluation. And if it is recognised that, even in its most rigorous forms, knowledge is not *wertfrei*, then the assumption of it as basis of a moral or legal deontology means the arbitrary setting up as a general rule of social behaviour what is in fact the product of conventions and decisions: *stat pro ratione voluntas*. It is more a matter of a 'deontological fallacy' which attributes prescriptive value to certain subjective evaluations than a 'naturalistic fallacy' in accordance with Hume's classic argument prohibiting the logical transition from 'being' to 'ought to be'.

From the point of view of a sociology of institutionalised beliefs ethical systems have their place in the complex of habits, practical rules, and symbolic codes on which individuals and groups base their social life: moral principles are part of the 'folklore' of cultural traditions which is transmitted from generation to generation. Ethics has therefore no ontological foundation: it does not reflect, as the traditional theory of natural law claims, an objective order of the world and society. There is no reason for ethics to be obeyed beyond the more or less free and conscious decision of individuals to make their own behaviour conform to certain rules. In other terms, deontological systems – ethics, law, politics – have no *Grundnorm* to make them, as in Kant or Kelsen, intrinsically valid and obligatory. And this is true not only of the ethical systems derived from mythical or religious cosmologies but also of the modern systems of rationalistic ethics justified by argumentation, such as contractualism and utilitarianism. [22] In particular no theory of justice is able to claim that it applies normatively to members of a social group fixed rules for the distribution of resources, privileges, and social status as being rules which are right, valid, and therefore obligatory.

This view gains strength from the recognition – with Neurath and against both classical and contemporary utilitarianism – of the diversity and heterogeneity of one individual's plans of life from another; also from the recognition that the expectations and disappointments of the individual members of a social group may only be compared ordinally, that no cardinal calculation is possible either to measure the 'utility functions' of an individual or, even more, to measure them in the aggregate. There are no criteria – neither that of *maximin* nor the neo-Bayesian one of the greatest expected utility – which can be taken as absolute rules for rationality of decision and maximisation of individual utility.[23] Moreover it would only be on the basis of an authority-decision that the position of an 'impartial observer' or original underwriter to a moral contract could be established in order to legitimise interpersonal comparisons of utility and decisions to maximise collective utility.[24]

The difficulties of the moralistic approach to the problem of collective choice become all the more serious when one recognises – with Hume and Neurath against Locke, Kant, and Rawls – that the contractualistic scheme is no more than a basis created for argument: it surreptitiously introduces the *sub specie aeternitatis* point of view of the noumenic individual, originally imagined free, rational, and equal, who gives an impartial or equitable judgement on the competition between different distributive theories and the underlying social conflict.

Beyond the moral metaphors of utilitarianism and contractualism one either supposes that the view-point of a universalistic ethics is shared by all the members of the social group – although in the complex modern societies of today such a hypothesis must seem wholly scholastic – or one recognises that a certain ethical point of view, along with the distributive criteria it intends to justify, is in fact imposed on individuals through the instruments of law and politics. There occurs, that is, a regulation by authority of the social conflict, over and above all ethical or 'rational' justification of political obligation.

From the ontologically and deontologically sceptical standpoint of reflexive epistemology it is not possible to prove 'scientifically' that one political system is preferable to another, for the reason that we do not have – in politics any more than in economics or physics – any solution to maximise the requirements of a universal rationality. This is notwithstanding the fact that, in the debate between democracy and autocracy, between liberalism and police state, between protection of individual rights and centralised planning, it is possible to argue epistemologically with some security for recognition of the partiality of all view-points and all forms of knowledge, and the contingency and relativity of all social expectations. An 'open' pluralistic system – one, that is, which is always alterable and reformable – is therefore 'rationally' preferable to totalistic planning regimes of the 'Platonic' type, even if it is not always easy to define what is meant by political pluralism.[25]

This does not remove the possibility that even a formally democratic regime may work against the interests of individuals, or minorities, or even majorities, in just the same way as a 'closed' despotic one, in so far as it may actually and permanently exclude their particular expectations from its own legitimised political complexity. And it is also in theory possible to argue, with Neurath, that no political system is able, in conditions of scarcity, to take account of the point of view of all members of the social group, nor is it able to avoid classing certain forms of behaviour as 'deviant'. Its 'openness' can never become total without running the danger of losing its own institutional stability.[26]

In this way, then, there are no reasons to follow Popper in maintaining that recourse to open conflict, civil disobedience, or 'revolution' should, even within a formally democratic regime, be, for 'scientific reasons', declared *a priori* irrational or immoral. In order to claim a complete ethico-political legitimacy, the controls of a democratic government should at all times be freely shared by the totality of citizens. Democracy would be unable to justify itself on the basis of any principle other than that of permanent and unanimous consensus: it could not establish itself either on the representative system or on the majority principle.

3. EPISTEMOLOGICAL PRIORITY OF THE SOCIAL SCIENCES AND SOCIAL COMPLEXITY

In the epistemology of the social sciences the reflexive approach suggests that significant revision is needed of the naturalistic and behaviouristic presuppositions on which contemporary 'social sciences' in Europe and the United States are based. Even in its more moderate variants, behaviourism applies a non-reflexive, and so dogmatic, reading of empiricism to the social sciences, in particular to sociology and political science. It makes the factual verifiability or falsifiability of scientific propositions into a condition of their scientificity, sets rigorously nomological standards of explanation and prediction of social phenomena, and demands an absolute undertaking of *Wertfreiheit* from the social scientist.

A reflexive reading of empiricism, on the other hand, takes account of the already high and still growing complexity of social phenomena and sees none of these methodological conditions as a requirement of the scientificity of the social disciplines. As Neurath was first of those in the empirical tradition to indicate, the complexity of social phenomena is such that a claim to work in sociology with the methods of naturalistic observation and quantitative measurement amounts simply to 'scientific suicide'. The high level of evolutive indeterminateness in human groups, their organisational structures and cultural forms, and their openness to a large number of deeply differentiated and 'improbable' possibilities renders the forms of social change in practice unpredictable. Even the behaviour of individuals within each group is equally complex and statistically unpredictable, on account not only of the dispersion and instability phenomena which characterise individual behaviour but also of the reflexive paradoxes of social prediction, involved as it always is in complex interactions between expectations and expectations of expectations.

If sociological research forms the preliminary condition of all possible 'rational' reconstructions of the languages and methods used by the scientific communities, then the epistemological problems of the social sciences cannot avoid preceding and conditioning the whole area of epistemological problems relative to the natural sciences. The scientist, in order to understand himself as scientist, has first to become anthropologist, sociologist, psychologist, and historian of himself and only subsequently 'logician' or 'physicist' or 'neurologist' of himself. This is the sense in which we can speak with confidence of the epistemological priority of the social sciences.

The level of 'rationality' of the natural sciences cannot therefore exceed that of the social sciences, nor does it fall to the natural sciences to prescribe a model of scientific explanation to serve as the condition of epistemological legality of historical, anthropological, or sociological research. In respect of positivistic reductionism the relationship has therefore been entirely overturned and the interdisciplinary consideration of science, at whatever level it is carried out – linguistic, methodological, or nomological – will not now be able to avoid dealing with the problem of 'social complexity'. It will have to take account of the fact that the objects of analysis itself – the social agents – have the singular characteristic of being themselves producers of symbolic systems: their forms of behaviour would remain wholly incomprensible to scientific observation if the observer attempted to leave out of consideration the motives, intentions, plans, and codes explicitly referred to by the verbal behaviour of the social agents and constantly alluded to also by their non-verbal behaviour. Thus even the epistemology of the natural sciences is forced to operate on the basis of the standards practised by the communities of social scientists: weak generalisations, incomplete explanations, bounded rationality.[27]

In this way one could maintain that the 'strong programme' of Barnes and Bloor fails to measure up to the important objectives it sets itself, since it claims to operate according to standards of 'maximal generality' and eschews, in the name of the scientificity of sociological research, all reference to the interpretative procedures of 'philosophy'. The sociology of science, in as much as it is a 'science of science', as Bloor argues in enunciating the principle of reflexivity in the 'strong programme', is a causalistic science like physiology and mechanics and it is as a result of this homogeneity of methods and principles that science can "scientifically know itself". [28] This is indeed a reductive development of both the 'holistic-reflexive' principle of Duhem-Neurath and the 'naturalised epistemology' of Quine.

If on the other hand it is maintained that the reflexivity of epistemological research cannot be concealed in the short-circuit of causalistic research but demands above all a process of self-reflection from the historico-social sciences, then the preliminary question becomes that of which methods are in effect practised by the communities of historians and social scientists. And it remains extremely difficult to show, even after the programmatic statements of the 'Edinburgh School', that this is a matter of causalistic methods.[29]

A reflexive epistemology of the social sciences will need to establish itself, thanks to a kind of epistemological feedback loop, as a 'sociology of sociology', and thus as a reflexive interpretation of what scientists do, say, and say they do when they do 'social science'. It will involve an attempt at self-comprehension which will start from the initial, purely stipulative and conjectural, acceptance of the language, theories, and empirical procedures practised by the communities of social researchers and will apply these categories to themselves in a recursive process of successive adjustments, new hypotheses, and new stipulations.

The significance, productiviness, and acceptability of this 'reflexive' strategy will depend not upon its foundations but upon its results: upon its ability to clarify languages and the stipulative procedures that legitimise them, to understand the 'paradigms' or 'styles of thought' underlying all 'encyclopedias' of social knowledge, to identify problems and objectives 'rationally'. It will depend not upon its own (in any case unattainable) *Wertfreiheit*, but upon its ability to harness the 'forces of love and hate' within the scope of a 'thinkful wishing' as far removed from ideological fanaticism as it is from Carnap's 'purity' of scientific method.

Only a social science conscious of its own language, its own methods, and its own aims – conscious, that is, of its own 'meaning' – would be in a position to study the languages, methods, and aims of science in its entirety. It would provide for the inception of a less schizophrenic, more 'wise', relationship between science and society, between manipulative power and ecological well-being, between technological growth and 'quality of life'.[30]

In this way Neurath's encyclopedic idea and inspired attempt to promote a unitary 'orchestration' of the language of science – even if not of course his own specific project – may possibly regain something of its importance and relevance. In spite of all the dangers implicit in this, it is evident that it would meet a widely felt demand for the interdisciplinary recovery of the 'meaning' of scientific research as a whole over and above its growing specialisation and impressive technological results. [31] This recovery now seems necessary in order to understand the reasons for and to evaluate the risks behind the increasing capacity to mould the forms of life and human knowledge which Galilean science has high-handedly acquired for itself in the course of this century, reaching the point where it has arrogated a kind of unassailable 'objectivity' and 'universality' in a no longer ontological but practico-evolutive sense.

In the context of a reflexive epistemology the encyclopedic idea would entail unitary consideration of the natural and social sciences. The classical distinction between explanatory and nomographical 'natural sciences' and understanding and ideographical 'social sciences' would lose the traditional meanings which it has acquired in the course of the widespread but not always productive debate between the theorists of *Erklären* and of *Verstehen*. Neither would the natural sciences be capable of being thought of as the exclusive domain of rigorous thought, where logic and mathematics provide the function of a grammatical guarantee of a perfectly clear and deductively binding communication, nor would the social sciences be capable of being defined as 'inexact sciences', the realm of undetermined conceptualisations, or, as Popper has put it, of ideologies and untestable methods. Such an 'encyclopedic' idea would start from the recognition that in all areas of knowledge scientific research must not only deal with a large number of paradoxes, anomalies, and exceptions but is also forced to operate in conditions of increasing complexity, as a result of the growth in complexity of the environment from which the spiral of human thought and knowledge arises.

With the demise of the 'universalistic illusion' which has fascinated classical science from Galileo down to (and including) Einstein, it is today physics itself which, at its more advanced frontiers such as thermodynamics, astrophysics, and quantum mechanics, is tending to become a 'science of complexity'. [32] It is rejecting all atemporal 'extraterritoriality' of scientific knowledge in respect of the historical experience of *homo sapiens*, the evolution of cultures, interaction with the environment, and the processes which the 'vortex of time' renders unpredictable and irreversible. In this way therefore it becomes amenable to a *nouvelle alliance* with the human sciences, because it takes account of the fact that, to take up Neurath's metaphor once more, it travels with them in open sea – or open sky – on the same boat in constant need of self-repair and restructuring, while all communication with the distant shore of certainty and homely reassuring truths has now long been cut off.

NOTES

[1] Cf.: H. Putnam, 'What is 'Realism'?', *Proceedings of the Aristotelian Society*, lxxvi (1975–76), p. 18 177; H. Putnam, 'Realism and Reason', in H. Putnam, *Meaning and the Moral Sciences*, London: Routledge & Kegan Paul, 1978; M. Hesse, *Revolutions and Reconstructions in the Philosophy of Science*, pp. xii–xiv; N. Rescher, *Scientific Realism. A Critical Reappraisal*, Dordrecht and Boston: D. Reidel, 1987.

[2] A particularly brilliant and radical expression of this attitude is to be found in R. Rorty's book, *Philosophy and the Mirror of Nature*; see particularly chapters 3 and 4, pp. 131–212.

[3] Cf.: J.R. Brown, 'Introduction: The Sociological Turn', in J. R. Brown (ed.), *Scientific Rationality: The Sociological Turn*, Dordrecht and Boston: D. Reidel, 1984,

pp. 3–26 (see also the bibliography, ibid., pp. 26–40); S.B. Barnes and D. Edge (eds.), *Science in Context: Readings in the Sociology of Science*, Boston: MIT Press, 1982.

[4] In my essay 'The Epistemological Status of the Theory of Autopoiesis and Its Applications to the Social Sciences', in G. Teubner (ed.), *Autopoiesis in Law and Society*, Berlin and New York: De Gruyter, in progress, I have tried to put forward some systematic arguments against the concept of 'experimental epistemology'.

[5] Cf. David M. Ricci, *The Tragedy of Political Science*, New Haven: Yale University Press, 1984; on the crisis of the behaviouristic social and political sciences see for instance: R.J. Bernstein, *The Restructuring of Social and Political Theory*, Philadelphia: University of Pennsylvania Press, 1976; R.F. Dallmayr, T.A. McCarthy (eds.), *Understanding Social Inquiry*; A. Ryan, *The Philosophy of Social Explanation*, Oxford: Oxford University Press, 1978.

[6] On the concept of 'social complexity' see: N. Luhmann, *Trust and Power*, Chichester and New York: Wiley, 1979, particularly pp. 4–9: N. Luhmann, *The Differentiation of Society*, New York: Columbia University Press, 1982.

[7] C. Cherniak includes 'computational complexity' within the limitations of the epistemic agent's rationality and views 'Neurath's boat' argument as a general limit to knowledge (*Minimal Rationality*, pp. 78–81, 107–9, 123–5).

[8] Cf.: W.V.O. Quine, *From a Logical Point of View*, pp. 42–6; W.V.O. Quine, 'Carnap and Logical Truth', in *New Readings in Philosophical Analysis*, ed. by H. Feigl, W. Sellars and K. Lehrer, pp. 95–109; W.V.O. Quine, 'On Carnap's View on Ontology', ibid, pp. 597–601. See also P. Lorenzen, *Constructive Philosophy*, pp. 3–29.

[9] In this sense also cf. late Wittgenstein in *Philosophical Investigations*, ed. by R. Rhees, Oxford: Basil Blackwell & Mott, 1958, pp. 25–6, 143. See also G.H. von Wright, *Wittgenstein*, Oxford: Basil Blackwell, 1982, pp. 165–82.

[10] For Kuhn's modified stance on this subject see: T.S. Kuhn, 'What are Scientific Revolutions?', in L.J. Daston, M. Heidelberger and L. Krüger (eds.), *The Probabilistic Revolution*, vol. 1: *Ideas in History*, Cambridge (Mass.): MIT Press, forthcoming; T.S. Kuhn, 'Possible Worlds in History of Science', paper presented at the Nobel Symposium, August 1986, to be published in *Possible Worlds in Humanities, Arts, and Sciences*, Berlin and New York: De Gruyter. See also: G. Gutting (ed.), *Paradigms and Revolutions*, Notre Dame (Ind.): University of Notre Dame Press, 1980; P. Hoyningen-Huene, *Die Wissenschaftsphilosophie Thomas S. Kuhns*, Braunschweig: Vieweg, 1989.

[11] Cf.: W.V.O. Quine, 'Truth by Convention', in *Readings in the Philosophy of Mathematics*, ed. by P. Benacerraf and H. Putnam, Cambridge and New York: Cambridge University Press, 1983; H. Putnam, 'The Analytic and the Synthetic', *Minnesota Studies in the Philosophy of Science*, III, ed. by H. Feigl and G. Maxwell, Minneapolis: University of Minnesota Press, 1962, pp. 358–97; H. Putnam, 'Is Logic Empirical?', *Boston Studies in the Philosophy of Science*, V, ed. by R.S. Cohen and M.W. Wartofsky, Dordrecht and Boston: D. Reidel, 1969, pp. 216–41.

[12] Cf. M. Hesse, 'Socializing Epistemology', in E. McMullin (ed.), *Construction and Constraint*, pp. 97–122. See also M. Hollis and S. Lukes, 'Introduction', in M. Hollis and S. Lukes (eds.), *Rationality and Relativism*, Oxford: Basil Blackwell, 1982, pp. 1–20.

[13] While recognising that Neurath's metaphor captures an essential truth, Lorenzen's

'constructive philosophy' is an attempt to find a methodological beginning of thought by breaking the linguistic and philosophical circularity which in his view includes both logical positivism and hermeneutics (cf. *Constructive Philosophy*, pp. 3–97). As far as Cherniak's interpretation of 'Neurath's boat argument' is concerned, it is not exactly clear to me whether he understands it, in Peirce's vein, as an argument for 'critical commonsensisms' (we must, by and large, accept our current scheme because if we try to undertake Descartes's method of universal doubt we are sunk) or as an argument for a reflexive philosophical stance (cf. *Minimal Rationality*, pp. 107–9).

[14] Cf. M. Hesse, *Revolutions and Reconstructions in the Philosophy of Science*, pp. vii-xx; see also N. Rescher, *The Limits of Science*, Berkeley: University of California Press, 1984, particularly pp. 60–94.

[15] Cf.: N. Luhmann, *Soziologische Aufklärung*, I, Opladen: Westdeutscher Verlag, 1970; N. Luhmann, *Politische Planung*, Opladen: Westdeutscher Verlag, 1971. On the epistemological aspects of Luhmann's theory of social complexity see my essays: 'Function, Meaning, Complexity. The Epistemological Premisses of Niklas Luhmann's 'Sociological Enlightenment'', *Philosophy of the Social Sciences, 16* (1986), no. 2; 'Reflexive Selbstbegründung der Soziologie und Autopoiesis. Über die epistemologischen Voraussetzungen der 'allgemeinen Theorie sozialer Systeme' Niklas Luhmanns', *Soziale Welt, 36* (1985), no. 4.

[16] Cf.: M. Hesse, 'An Inductive Logic of Theories', in *Minnesota Studies in the Philosophy of Science*, IV, ed. by M. Radner and S. Winokour, Minneapolis: University of Minnesota Press, 1970, pp. 164–80; M. Hesse, 'Consilience of Inductions', in *The Problem of Inductive Logic*, ed. by I. Lakatos, 1968, pp. 232–46; M. Hesse, 'Bayesian Methods and the Initial Probabilities of Theories', in *Minnesota Studies in the Philosophy of Science*, VI, ed. by G. Maxwell and R.M. Anderson, Minneapolis: University of Minnesota Press, 1975, pp. 50–105.

[17] Cf.: W.C. Salmon, 'Statistical Explanation', in R.G. Colodny (ed.), *The Nature and Function of Scientific Theories*, Pittsburgh: University of Pittsburgh Press, 1970, pp. 173–231; W.C. Salmon, 'The Foundations of Scientific Inference', in R.G. Colodny (ed.), *Mind and Cosmos*, pp. 135–75; W.C. Salmon, R.C. Jeffrey and J.G. Greeno, *Statistical Explanation and Statistical Relevance*, Pittsburgh: University of Pittsburgh Press, 1979.

[18] Cf.: M. Brodbeck, 'Explanation, Prediction and 'Imperfect' Knowledge', in M. Brodbeck, *Readings in the Philosophy of the Social Sciences*, pp. 363–97; M. Scriven, 'The Key Property of Physical Laws: Inaccuracy', in *Current Issues in the Philosophy of Science*, ed. by H. Feigl and G. Maxwell, New York: Holt, Rinehart and Winston, 1961, pp. 91–104; M. Scriven, 'Explanations, Predictions and Laws', in *Readings in the Philosophy of Science*, ed. by B.A. Brody, Englewood Cliff (NJ.): Prentice-Hall, 1970, pp. 88–104.

[19] Cf.: M. Hesse, *Models and Analogies in Science*, Notre Dame (Ind.): University of Notre Dame Press, 1966; M. Hesse, 'Positivism and the Logic of Scientific Theories', in P. Achinstein and S.F. Barker (eds.), *The Legacy of Logical Positivism*, pp. 85–114.

[20] Cf. L. Fleck, *Genesis and Development of a Scientific Fact*, p. 21. See also: L. Fleck, 'The Problem of Epistemology', in R.S. Cohen and T. Schnelle (eds.), *Cognition and Fact. Materials on Ludwik Fleck*, particularly pp. 79–81, 97–105; L. Fleck, 'Problems of the Science of Science', ibid., pp. 113–27.

[21] Cf. M. Bunge, *Causality. The Place of the Causal Principle in Modern Science*,

Cleveland: World Pub. Co., 1963. See also M.A. Arbib and M.B. Hesse, *The Construction of Reality*, Cambridge: Cambridge University Press, 1986, pp. 197–221.
[22] Cf. J. Elster, *Ulysses and the Sirens*, Cambridge: Cambridge University Press, 1979, pp. 112–56.
[23] See, contra, J.C. Harsanyi, *Rational Behavior and Bargaining Equilibrium in Games and Social Situations*, Cambridge: Cambridge University Press, 1976.
[24] In this sense too both Sen's theorem of the impossibility of a Paretian liberal, and Arrow's classical theorem of impossibility may be interpreted; cf.: A.K. Sen, 'The Impossibility of a Paretian Liberal', *Journal of Political Economy*, *78* (1970), pp. 152–7; K.J. Arrow, *Social Choice and Individual Values*, New Haven: Yale University Press, 1963.
[25] Cf.: N. Bobbio, *The Future of Democracy*, Cambridge: Polity Press, 1987; N. Bobbio, *Which Socialism?*, Cambridge: Polity Press, 1987.
[26] Cf.: P. Bachrach, S.M. Baratz, *Power and Poverty*, New York: Oxford University Press, 1970; E.E. Schattschneider, *The Semi-Sovereign People: A Realist's View of Democracy in America*, New York: Holt, Rinehart & Winston, 1960.
[27] Cf. H.A. Simon, *The Sciences of the Artificial*, Cambridge (Mass.): MIT Press, 1969.
[28] Cf. D. Bloor, *Knowledge and Social Imagery*, London: Routledge and Kegan Paul, 1976, p. 40. See also J.R. Brown, 'Introduction: The Sociological Turn', in J.R. Brown (ed.), *Scientific Rationality: The Sociological Turn*, pp. 10–11.
[29] Cf.: W.H. Newton-Smith, *The Rationality of Science*, Boston and London: Routledge & Kegan Paul, 1981, p. 264; L. Laudan, 'The Pseudo-science of Science?', in J.R. Brown (ed.), *Scientific Rationality: The Sociological Turn*, pp. 48–53.
[30] Cf.: W. Schäfer (ed.), *Finalization in Science*, Dordrecht and Boston: D. Reidel, 1983; N. Maxwell, *From Knowledge to Wisdom*, Oxford: Basil Blackwell, 1984.
[31] Cf. W. Bechtel, 'The Nature of Scientific Integration' in W. Bechtel (ed.), *Integrating Scientific Disciplines*, Dordrecht: Martinus Nijhoff Publishers, 1986, pp. 3–52.
[32] Cf. I. Prigogine and I. Stengers, *La Nouvelle alliance. Métamorphose de la science*, Paris: Gallimard, 1979, English trans. *Order out of Chaos*, Boulder and London: Shambhala, 1984, particularly pp. 103–209.

LIST OF OTTO NEURATH'S CITED WORKS*

[1910] 'Zur Theorie der Sozialwissenschaften', *Jahrbuch für Gesetzgebung, Verwaltung und Volkswirtschaft im Deutschen Reich, 34*, pp. 37–67.
GpmS, pp. 23–46. (ON 16)

[1912] 'Das Problem des Lustmaximums', *Jahrbuch der Philosophischen Gesellschaft an der Universität zu Wien 1912*, pp. 89–100.
GpmS, pp. 47–55. (ON 50)

[1913] 'Die Verirrten des Cartesius und das Auxiliarmotiv: Zur Psychologie des Entschlusses', *Jahrbuch der philosophischen Gesellschaft an der Universität zu Wien 1913*, pp. 45–59.
GpmS, pp. 57–67. *PP*, pp. 1–12. (ON 62)

[1915] 'Prinzipielles zur Geschichte der Optik', *Archiv für Geschichte der Naturwissenschaften und der Technik, 5*, pp. 371–89.
GpmS, pp. 71–84.
ES, pp. 101–12. (ON 81)

[1916] 'Zur Klassifikation von Hypothesensystemen. (Mit besonderer Berücksichtigung der Optik)', *Jahrbuch der philosophischen Gesellschaft an der Universität zu Wien 1914 und 1915*, pp. 371–89.
GpmS, pp. 85–101.
PP, pp. 13–31. (ON 82)

[1917] 'Das Begriffsgebäude der Wirtschaftslehre und seine Grundlagen', *Zeitschrift für die gesamte Staatswissenschaft, 73*, pp. 484–520.
GpmS, pp. 103–29. (ON 91)

[1918] 'Josef Popper-Linkeus: seine Bedeutung als Zeitgenosse', *Neues Frauenleben, 20*, n. 3, pp. 33–8.
GpmS, pp. 131–6. (ON 96)

[1919] 'Utopien', *Wirtschaft und Lebensordnung*, 1, no. 1. pp. 5–6.
GpmS, pp. 137–8. (ON 101)

* With 'ON' followed by a number and put within brackets reference is made to the 'List of works by Otto Neurath' (which contains a total of 277 entries) compiled by Marie Neurath in *ES*, pp. 440–59. When the number is followed by a letter, as well as for numbers 278, 279 and 280, reference is made to the 'Supplementary list of works by Otto Neurath', in *PP*, pp. 255–8.

[1919a] *Durch die Kriegswirtschaft zur Naturalwirtschaft*, München: Georg D. W. Callwey.
ES, pp. 123–57. (ON 114)

[1921] *Anti-Spengler*, München: Georg D.W. Callwey.
GpmS, pp. 139–96.
ES, pp. 158–213. (ON 129)

[1928] *Lebensgestaltung und Klassenkampf*, Berlin: E. Laub.
GpmS, pp. 227–93.
ES, pp. 249–98. (ON 172)

[1928a] Review of R. Carnap's books: *Der logische Aufbau der Welt* and *Scheinprobleme in der Philosophie*, in *Der Kampf*, *21*, pp. 624–6.
GpmS, pp. 295–7. (ON 174)

[1929] *Wissenschaftliche Weltauffassung: Der Wiener Kreis* (with R. Carnap and H. Hahn), Vienna: Artur Wolf.
GpmS, pp. 299–336.
ES, pp. 299–318. (ON 179)

[1930] 'Bürgerlicher Marxismus', *Der Kampf*, *23*, pp. 227–32.
GpmS, pp. 349–56. (ON 184)

[1930a] Review of W. Sombart, *Die drei Nazionalökonomien*, in *Der Kampf*, *23*, pp. 300–2.
GpmS, pp. 357–61. (ON 185e)

[1930/31] 'Wege der wissenschaftlichen Weltauffassung', *Erkenntnis*, *1*, pp. 106–25.
GpmS, pp. 371–85.
PP, pp. 32–47. (ON 187)

[1931] 'Marxismus eines Jesuiten' (review of I.B. Kraus, *Scholastik, Puritanismus und Kapitalismus*), *Der Kampf*, *24*, pp. 271–4.
GpmS, pp. 401–5 (ON 195)

[1931a] 'Weltanschauung und Marxismus', *Der Kampf*, *24*, pp. 447–51.
GpmS, pp. 407–12. (ON 196)

[1931b] 'Magie und Technik', *Erkenntnis*, *2*, pp. 82–4.
GpmS, pp. 529–31. (ON 196e)

[1931c] 'Physicalism: The Philosophy of the Viennese Circle', *The Monist*, *41*, pp. 618–23.
GpmS, pp. 413–6. *PP*, pp. 48–51. (ON 197)

[1931d] 'Physikalismus', *Scientia, 50*, pp. 297–303.
GpmS, pp. 417–21.
PP, pp. 52–7. (ON 198)

[1931e] *Empirische Soziologie*, Wien: Julius Springer.
GpmS, pp. 423–527.
ES, pp. 319–421. (ON 201)

[1931f] 'Soziologie im Physikalismus', *Erkenntnis, 2*, pp. 393–431.
GpmS, pp. 533–62.
PP, pp. 58–90. (ON 202)

[1932] 'Sozialbehaviorismus', *Soziologus 8*, pp. 281–88.
GpmS, pp. 563–9. (ON 207)

[1932a] 'Die 'Philosophie' im Kampf gegen den Fortschritt der Wissenschaft', *Der Kampf, 25*, pp. 385–89.
GpmS, pp. 571–6. (ON 209)

[1932/33] 'Protokollsätze', *Erkenntnis, 3*, pp. 204–14
GpmS, pp. 577–85.
PP, pp. 91–9. (ON 210)

[1933] *Einheitswissenschaft und Psychologie*, Wien: Gerold & Co.
GpmS, pp. 587–610. (ON 216)

[1934] 'Radikaler Physikalismus und 'Wirkliche Welt'', *Erkenntnis, 4*, pp. 346–62.
GpmS, pp. 611–523.
PP, pp. 100–14. (ON 218)

[1935] 'Einheit der Wissenschaft als Aufgabe', *Erkenntnis, 5*, pp. 16–22.
GpmS, pp. 625–30.
PP, pp. 115–20. (ON 219)

[1935a] 'Pseudorationalismus der Falsifikation', *Erkenntnis, 5*, pp. 353–65.
GpmS, pp. 635–44.
PP, pp. 121–31. (ON 220)

[1935b] 'Zur Induktionsfrage', *Erkenntnis, 5*, pp. 173–4.
GpmS, pp. 631–2. (ON 223a)

[1935c] 'Erster Internationaler Kongreß für Einheit der Wissenschaft in Paris 1935', *Erkenntnis, 5*, pp. 377–406.
GpmS, pp. 649–71. (ON 223c)

[1935d] Review of A. Ross, *Kritik der sogenannten praktischen Erkenntnis*, in

Erkenntnis, *5*, pp. 366–7.
GpmS, pp. 645–7. (ON 223d)

[1936] *Le développement du Cercle de Vienne et l'avenir de l'empirisme logique*, Paris: Hermann & Cie.
GpmS, pp. 673–702. (ON 224)

[1936a] 'Einzelwissenschaften, Einheitswissenschaft, Pseudorationalismus', in *Actes du Congrès internationale de philosophie scientifique, Sorbonne, Paris 1935*, I, *Philosophie scientifique et empirisme logique*, in *Actualités Scientifique et Industrielles*, no. 388, Paris, Hermann & Cie, pp. 57–64.
GpmS, pp. 703–9.
PP, pp. 132–8. (ON 225)

[1936b] 'Mensch und Gesellschaft in der Wissenschaft', in *Actes du Congrès internationale de philosophie scientifique, Sorbonne, Paris 1935*, II, *L'unité de la science*, ibid., no. 389, pp. 32–40.
GpmS, pp. 711–7. (ON 226)

[1936c] 'Un encyclopédie internationale de la science unitaire', in *Actes du Congrès internationale de philosophie scientifique, Sorbonne, Paris 1935*, II, *L'unité de la science*, ibid., no. 389, pp. 54–59.
GpmS, pp. 719–24.
PP, pp. 139–44. (ON 227)

[1936d] 'L'encyclopédie comme 'modèle'', *Revue de Synthèse*, *12*, No. 2, pp. 187–201.
GpmS, pp. 725–38.
PP, pp. 145–58. (ON 228)

[1936e] 'Den Logiska Empirismen och Wienerkretsen', *Theoria*, *2*, pp. 72–82 (Goeteborg).
GpmS, ('Der Logische Empirismus und der Wiener Kreis') pp. 739–47. (ON 229)

[1936f] 'Physikalismus und Erkenntnisforschung', *Theoria*, *2*, no. 1, pp. 97–105, no. 2, pp. 234–6.
GpmS, pp. 749–59.
PP, pp. 159–71. (ON 230)

[1936g] 'Einheitswissenschaft', *Actes du Huitième Congrès international de philosophie à Prague, 2–7 septembre 1934*, Prag: Orbis, pp. 139–41.
GpmS, pp. 761–3. (ON 232)

[1936h] 'Soziologische Prognosen', *Erkenntnis*, *6*, pp. 398–405.
GpmS, pp. 771–6. (ON 234)

[1937] 'Inventory of the Standard of Living', *Zeitschrift für Sozialforschung*, *6*, no. 1, pp. 140–51. (ON 237)

[1937a] 'Unified Science and Its Encyclopedia', *Philosophy of Science*, *4*, pp. 265–77.
GpmS, pp. 777–86.
PP, pp. 172–82. (ON 239)

[1937b] 'Prognosen und Terminologie in Physik, Biologie, Soziologie', in *Travaux du IXe Congrès international de philosophie*, IV, *L'unité de la science: la méthode et les méthodes*, Ie partie, *Actualités Scientifique et Industrielles*, no. 533, Paris: Hermann & Cie, pp. 77–85.
GpmS, pp. 787–94. (ON 241)

[1937c] 'Die neue Enzyklopädie des wissenschaftlichen Empirismus', *Scientia*, *62*, pp. 309–20.
GpmS, pp. 801–11.
PP, pp. 189–99. (ON 243)

[1937/38] 'The Departmentalisation of Unified Science', *Journal of Unified Science* (*Erkenntnis*), *7*, pp. 240–6.
GpmS, pp. 813–19.
PP, pp. 200–5. (ON 244)

[1937/38a] 'Zu den Vorträgen von Black, Kokoszynska, Williams', *Journal of Unified Science* (*Erkenntnis*), *7*, pp. 371–74.
GpmS, pp. 821–3.
PP, pp. 206–8. (ON 245)

[1938] 'Eenheidswetenschap als empiristische Synthese', *Synthese*, *3*, no. 1, pp. 10–7.
GpmS ('Einheitswissenschaft als empiristische Synthese'), pp. 825–31. (ON 246)

[1938a] 'Weten en Zijn', *Algemeen Nederlands Tijdschrift voor Wijsbegeerte en Psychologie*, *31*, no. 5, pp. 230–9.
GpmS ('Wissen und Sein'), pp. 833–43. (ON 247)

[1938b] 'Die wissenschaftliche Sprache', in *Unity of Science Forum*, Supplement to *Synthese*, *3*, no. 8, pp. 1–8.
GpmS, pp. 845–51. (ON 248)

[1938c] 'Die neue Enzyklopädie', in O. Neurath, et al., *Zur Enzyklopädie der Einheitswissenschaft*, 6, pp. 6–16, Den Haag: Van Stockum & Zoon.
GpmS, pp. 863–71. (ON 251)

[1938d] 'Unified Science as Encyclopedic Integration', in O. Neurath, et al.,

Encyclopedia and Unified Science, in *International Encyclopedia of Unified Science*, vol. 1, no. 1, pp. 1–27, Chicago: The University of Chicago Press.
GpmS, pp. 873–94. (ON 252)

[1939/40] 'The Social Sciences and Unified Science', *The Journal of Unified Science (Erkenntnis)*, reprints made from the proofs, *9*, pp. 430–3.
GpmS, pp. 895–9.
PP, pp. 209–12. (ON 255)

[1941] 'Universal Jargon and Terminology', *Proceedings of the Aristotelian Society*, New Series, *41*, pp. 127–48.
GpmS, pp. 901–18.
PP, pp. 213–29. (ON 256)

[1941a] 'The Danger of a Careless Terminology', *The New Era*, *22*, no. 7, pp. 145–50.
GpmS, pp. 919–24. (ON 257)

[1942] 'International Planning for Freedom', *The New Commonwealth Quarterly*, April, pp. 281–92; July, pp. 23–8.
ES, pp. 421–40. (ON 258)

[1943] 'Planning or Managerial Revolution?', *The New Commonwealth Quarterly*, April, pp. 148–54. (ON 260)

[1944] 'Ways of Life in a World Community', *The London Quarterly of World Affairs*, July, pp. 29–32. (ON 264)

[1944a] *Foundations of the Social Sciences*, Chicago: The University of Chicago Press. (ON 265)

[1944b] 'Nazi Text-books and the Future', *The Journal of Education*, *76*, I, no. 904, pp. 521–2; II, no. 905, pp. 574–6 (with J.A. Lauwerys). (ON 266)

[1945] Review of F.A. Hayek, *The Road to Serfdom*, in *The London Quarterly of World Affairs*, January, pp. 121–2. (ON 267)

[1945a] 'Plato's Republic and German Education', *The Journal of Education*, *77*, no. 907, pp. 57–9; no. 910, pp. 222–4; no. 913, p. 394 (with J.A. Lauwerys). (ON 268)

[1945b] 'Germany's Education and Democracy', *The Journal of Education*, *77*, no. 912, pp. 370–1. (ON 269)

[1945/46] 'The Orchestration of the Sciences by the Encyclopedism of Logical Empiricism', *Philosophy and Phenomenological Research*, *6*, no. 4, pp.

496–508.
GpmS, pp. 997–1009.
PP, pp. 230–42. (ON 272)

[1946a] 'Prediction and Induction', (compiled from notes by M. Neurath), *Analisi*, *1*, no. 3, pp. 1–6.
GpmS, pp. 1015–8.
PP, pp. 243–6. (ON 274)

[1973] *Empiricism and Sociology*, ed. by M. Neurath and R.S. Cohen, Dordrecht and Boston: D. Reidel. (ON 277)

[1979] *Wissenschaftliche Weltauffassung, Sozialismus und Logischer Empirismus*, ed. by R. Hegselmann, Frankfurt a.M.: Suhrkamp. (ON 278)

[1981] *Gesammelte philosophische und methodologische Schriften*, ed. by R. Haller and H. Rutte, Wien: Hölder-Pichler-Tempsky Verlag. (ON 279)

[1982] *Philosophical Papers 1913–1946*, ed. by R.S. Cohen and M. Neurath, Dordrecht and Boston: D. Reidel. (ON 280)

[1987] (ed.), *Unified Science*, English translation of the series of monographs *Zur Enzyklopädie der Einheitswissenschaft* originally edited by Otto Neurath, ed. by B.F. McGuiness and H. Kaal, Dordrecht and Boston: D. Reidel. (cf. ON 251)

[1989] *Economic Writings*, ed. by R.S. Cohen and M. Neurath, Dordrecht and Boston: D. Reidel. (forthcoming)

META-BIBLIOGRAPHICAL NOTE

1. A 'List of works by Otto Neurath' has been compiled by Marie Neurath and published as Chapter 12 in *ES*, pp. 441–59. R. Haller and E. Rutte, editors of *GpmS*, inform us (see 'Vorvort', p. xv) that their edition of Neurath's works includes 14 writings which had not been listed by Marie Neurath in her bibliography. More recently in *PP*, edited by Robert S. Cohen and Marie Neurath, there has appeared a 'Supplementary List of Works by Otto Neurath' (pp. 255–8), along with a list of Neurath's writings originally published in English or translated into English.

2. For the secondary bibliography on Neurath's social, philosophical and methodological thought see the following works: G. Statera, *Logica, linguaggio e sociologia. Studio su Otto Neurath e il neopositivismo*, pp. 189–92; F. Barone, 'Introduzione' to O. Neurath, *Il Circolo di Vienna e l'avvenire dell'empirismo logico*, pp. 9–43; K. Fleck, *Otto Neurath. Eine biographische und systematische Untersuchung*, pp. 142–50; E. Nemeth, *Otto Neurath und der Wiener Kreis*, pp. 211–3.

3. For the literature on the Vienna Circle and Logical Positivism the following compilations are still useful: the *Bibliography of Logical Positivism*, edited by A.J. Ayer in A.J. Ayer (ed.), *Logical Positivism*, pp. 381–446; the bibliographical appendix in F. Barone, *Il neopositivismo logico*, pp. 651–712. An important source of bibliographical information exists also in *Erkenntnis*, particularly *1* (1930–31), pp. 315–39, and *5* (1935), pp. 185–203, 409–28. See also J. Joergensen, *The Development of Logical Empiricism*, passim.

4. For a first bibliographical approach to the devolopment and the crisis of empiricism over the last forty years and to the relationship between empiricism and the social sciences see the bibliographical appendixes in F. Suppe (ed.), *The Structure of Scientific Theories*, pp. 731–67, and in M. Brodbeck (ed.), *Readings in the Philosophy of the Social Sciences*, pp. 737–68.

SUBJECT INDEX

AUTHOR INDEX